Random Fatigue:

From Data to Theory

Random Fatigue:
From Data to Theory

K. Sobczyk

Institute of Fundamental Technological Research
Polish Academy of Sciences
Warsaw, Poland

B. F. Spencer, Jr.

Department of Civil Engineering
University of Notre Dame
Notre Dame, Indiana, USA

ACADEMIC PRESS, INC.
Harcourt Brace Jovanovich, Publishers
Boston San Diego New York
London Sydney Tokyo Toronto

This book is printed on acid-free paper. ♾

ACADEMIC PRESS, INC.
1250 Sixth Avenue, San Diego, CA 92101

United Kingdom Edition published by
ACADEMIC PRESS LIMITED
24–28 Oval Road, London NW1 7DX

Library of Congress Cataloging-in-Publication Data

Sobczyk, Kazimierz.
Random fatigue: from data to theory / K. Sobczyk, B. F. Spencer, Jr.
p. cm.
Includes bibliographical references (p.) and index.
ISBN 0-12-654225-2
1. Materials—Fatigue. 2. Fracture mechanics. I. Spencer, B. F. (Billie Floyd), 1959- . II. Title.
TA418.38.S63 1992
620.1′126—dc20 91-33269
CIP

Printed in the United States of America

92 93 94 9 8 7 6 5 4 3 2 1

Contents

Preface

This book presents a concise and unified exposition of the existing methods of modelling and analysis of fatigue fracture of engineering materials. The most notable approaches to construction of stochastic models of fatigue processes are expounded with emphasis on their methodical consistency and potential usefulness in engineering practice.

The subject of this monograph — random fatigue — is of a specific nature. From an applied perspective, fatigue is concerned in an essential way with mechanics, metallurgy, and structural/mechanical reliability. On the other hand, experimental data from various laboratories constitute a basic source of information about fatigue of real engineering materials. In this situation, important problems arising are concerned with appropriate representation of the information contained in dispersed fatigue data in the language of mathematics. Because of the inherent randomness found in fatigue data, stochastic modelling of the fatigue process is both natural and necessary. Today, the problems associated with the theory of random fatigue are perceived as important and challenging tasks in mechanics and applied stochastics.

The study of random fatigue is now passing through a stage of rapid development. Due to this fact, it is difficult to accomplish a fully satisfactory treatment of the subject. Our intention has been to reflect the existing results as well as to show possible directions for future research. We hope that this monograph will also have an inspiring influence on the development of new types of fatigue experiments — designed and performed according to the random nature of fatigue.

This book is the result of our scientific interest and research in stochastic modelling of fatigue. It was written during the 1990-91 academic year when K. Sobczyk was the Melchor Visiting Professor in Engineering at the University of Notre Dame. Using this occasion, the authors wish to express their cordial thanks to the University of Notre Dame for making this visit possible and for providing excellent conditions for academic endeavors.

K. Sobczyk
B. F. Spencer, Jr.

Introduction:

From Data to Theory

For many years, fatigue has been a significant and difficult problem for engineers, especially for those who design structures such as aircraft, bridges, pressure vessels, cranes, etc. Fatigue of engineering materials is commonly regarded as an important deterioration process and a principal mode of failure for various structural and mechanical systems. Though no exact figures are available, it is believed that between 50 and 90 percent of all mechanical failures in metallic structures are fatigue-related. However, fatigue is a complicated process that is difficult to accurately describe or model. Still, fatigue damage assessment must be made for the purpose of engineering design. The need to accurately assess fatigue damage and fatigue life is particularly evident in the aircraft industry, where highly reliable structures are required, and fatigue is known to be a major problem. In fact, a relative numerical ranking of incidents of structural deficiencies in military aircraft placed fatigue cracking first over corrosion, maintenance/manufacturing damage, and other reported incidents. As a result, many research efforts on fatigue damage accumulation have been undertaken, and significant advances have been made toward the understanding of the fatigue behavior of engineering materials. These studies have also shed light on the physical nature of the fatigue phenomenon.

As one might expect, the fatigue process is a complex phenomenon that can be considered from several different viewpoints. For example, researchers in *physics* have been interested in the true physical (e.g., atomic, molecular) mechanisms of fatigue deterioration on the microscopic level, which are responsible for macroscopic effects (e.g., kinetic theory of fracture, or dislocation theory approaches for formulating a type of fatigue crack growth equation). While these efforts are of great importance, they cannot yet give a basis for the consistent and quantitative treatment of microscopic fatigue and fracture as it relates to the macroscopic phenomena

Fatigue is obviously a subject of *mechanics*. For mechanical engineers, the questions why, when, and where fatigue might occur, as well as its impact on the overall stress distribution in a structure, are of prime concern. Fatigue is an important issue in *metallurgy,* where manufacturers wish to produce structural elements that are resistant to fatigue deterioration. The metallurgist is concerned with basic factors such as alloy composition, heat treatment, and mechanical working when attempting to produce materials possessing high fatigue resistance.

In the design of engineering structures such as aircraft, bridges, pressure vessels, etc., fatigue is a significant problem with which a designer must reckon. Because many of the design parameters associated with fatigue possess considerable uncertainty, one comes to the problems commonly considered within reliability theory. For structures required to perform safely in the presence of uncertainty during their entire service life, it is necessary to have an appropriate probabilistic measure of ultimate fatigue failure. For this reason, the problems of *fatigue reliability* of structures (subjected to complex time-varying loading) are of great interest in recent research.

In spite of the problems mentioned before (physics of fatigue, mechanics of fatigue fracture, fatigue metallurgy, fatigue design) there exists an important issue that has significantly impacted the direction of recent research on fatigue. This is the fact that even today most of our basic knowledge on fatigue behavior comes from experiments. The experimental data from engineering laboratories constitute a basic source of information about the fatigue behavior of materials subjected to various loading conditions. In this situation, an important problem arising is: How can the information contained in fatigue data be represented consistently in the language of mathematics? This indicates a need for *mathematical modelling of fatigue*. However, experimental test results, as well as field data, indicate that the fatigue process in real materials involves considerable random variability. This variability depends upon many factors, such as material properties, type of loading, environmental conditions, etc. The random nature of fatigue is most obvious if a structure is subjected to randomly varying loading; but even in tightly controlled laboratory conditions under deterministic cyclic loading, results obtained show considerable statistical dispersion. (This will be shown in detail at the end of Chapter I.) Thus, because of the inherent randomness in fatigue data, *stochastic modelling* is both appropriate and necessary. This view has been widely accepted, and consequently, the problems associated with stochastic theory of fatigue are currently perceived as important and challenging tasks in mechanics and applied stochastics.

It would be extremely desirable to formulate a stochastic theory for the fatigue behavior of materials that deals with all of the physical and chemical processes on micro-scale and portrays the observed macroscopic characteristics of the fatigue process. Such modelling — truly rooted in the physics of random fatigue — does not currently seem to be possible. While the existing physical theories (e.g., thermodynamics, statistical physics) are helpful in qualitatively explaining fatigue behavior, they cannot yet give a basis for the *micro–macro* modelling of the fatigue process and for obtaining results of interest in engineering. In view of these difficulties, it is rational and important to formulate *macroscopic (phenomenological) stochastic theory* in order to recognize regularities in dispersed fatigue data and to provide a consistent basis for prediction of fatigue behavior for purposes of safety and reliability estimation.

* * *

General questions associated with modelling of real phenomena and with the mutual relationship between reality and theory are the subject of the methodology of empirical sciences and the philosophy of science. Although it is not our purpose here to detail this subject, it seems to be worth mentioning at least some basic questions that come to mind when one intends to build a mathematical (in particular, stochastic) model of a real process on the basis of information contained in empirical data. Some of the very first questions one must consider are:

- What is really meant by the statement that a certain system of mathematical symbols and relations is a representation of a real phenomenon?
- When are we justified in accepting that the conclusions arising logically from the model constitute true statements about reality?
- If different models of the same phenomenon are built, when can they be regarded as equivalent?
- Data are usually not complete; should one require a model to be completely adequate?

Most often the abstract or mathematical models of physical phenomena characterize these phenomena in a simplified way; the degree of this simplification depends upon the particular objective in modelling or, more often, on our (insufficient) knowledge of the phenomena in question.

A *model* is understood here as a mathematical representation of the essential aspects of the physical phenomenon (fatigue accumulation) that provides information about the phenomenon in a usable form. If the model is too complex, its usefulness becomes questionable. Therefore, relative simplicity seems to be an important feature in model construction.

An important issue in mathematical modelling is model validation through comparison with empirical data. It seems that according to common opinion, agreement with empirical data is not only necessary, but also sufficient for the acceptance of a scientific theory. This comes from the conviction that theories (or models) are just data summaries, or at worst, codifications of data and some slight extrapolation from them. Especially for practitioners, agreement with data seems to be the highest court of appeal. Of course, there is no question that agreement with experimental data must play a principal role in building a theory of any real phenomenon. However, one should also keep in mind that the mutual relationship between experiments and theory is complex, and that agreement with data cannot be the only criterion for the acceptance of a new theory.

While agreement between theory and experiments (under a given set of conditions) can confirm a theory, it does not point with certainty to its truth. It may happen that both theory and data are sloppy, and that compensating errors have entered into both. Neither can disagreement between theory and experience always be interpreted as a clear refutation of the theory. Of course, this is not to say that theories of real phenomena are unverifiable, but rather that the process of their empirical validation may be very complex and roundabout. Franklin writes in his book:[1] "Life is sometimes difficult on the frontier. This is true for both experiments and theory. Difficulty does not, however, mean impossibility."

There exist many factors that may affect experiments themselves and their resulting outcomes. For example, it is not always entirely clear whether a specific experimental apparatus is really capable of producing valid results. Usually one of the arts of an experimenter is to modify and adjust the apparatus until it does. One might also worry about the effects of the presuppositions of the experimenter on his experimental results; such presuppositions may cause an experimenter to overlook unexpected results. History has provided many cases in which experimenters have found desired results, even when the evidence was lacking.[2,3]

---

1. Franklin, A., *Experiments, Right or Wrong*, Cambridge University Press, Cambridge, Massachusetts, 1990.

2. Nye, M. J., "An Episode in the History and Psychology of Science," *Historical Studies in the Physical Sciences* **11**, 125–156 (1980).

3. Brush, S., "A History of Random Processes, I. Brownian Motion from Brown to Perrin," *Archive for the History of Exact Sciences* **5**, 5 (1968).

An empirical test of the mathematical model consists of confrontation of some logical consequences or predictions of the model with information obtained via observations or experiments. However, to consider an empirical test valid, the experiments have to be properly designed and performed, and the results should be carefully and correctly elaborated and interpreted. In the last century, Maxwell[4] wrote that when starting to test a theoretical hypothesis concerning physical reality, one should not rush into the laboratory, but rather further pursue some theoretical work: "The verification of the laws is affected by a theoretical investigation of the conditions under which certain quantities can be most accurately measured, followed by an experimental realization of these conditions, and actual measurements of the quantities."

The statement just cited emphasizes the fact that theoretical work must play a role in experimental design of empirical tests. Experimental design usually involves further hypotheses (i.e., in addition to the model) concerning the links between quantities in the theory and the observables. It often happens that the physical meaning of theoretical quantities and the corresponding observables are not the same (e.g., tests of a model for fatigue crack growth can be realized by measurement of striations on the fracture surface, or by the measurement of the parameters of acoustic emission).

In addition to experiments themselves, the analysis of data, including proper statistical inference, requires special care and knowledge. Improper data handling can significantly affect the reliability of the final results.

The complex and often inconclusive character of experimental testing of theoretical models can enhance the value of other features of the proposed models. The features that we have in mind are known as *metatheoretical* and *intertheoretical consistency*.[5] The first one means that the form and content of the model must be internally consistent (i.e., logically well-built, empirically testable, etc.). The second feature means that a new model should be compatible with other previously accepted theories (i.e., the model predictions cannot go against justified, verified, and commonly accepted beliefs).

Therefore, a model of physical phenomenon must be *reasonable* and *likely*, logically consistent, compatible with existing knowledge, and empirically testable. A new model possessing such attributes should then undergo a proper empirical verification. Such a model is deemed to be

4. Maxwell, J. C., "Remarks on the Mathematical Classification of Physical Quantities," *Proceedings of the London Mathematical Society* **3**, 224 (1871).

5. Bunge, M., "Theory Meets Experience," in *Mind, Science and History* (H. E. Kiefer, and M. K. Munitz, eds.), State University of New York, Albany, New York 1970

acceptable if it is consistent with empirical data and predicts correctly the evolution of the phenomenon.

* * *

Early studies devoted to probabilistic treatment of fatigue were mainly concerned with the statistics of dispersed fatigue data and application of the results to estimating the fatigue reliability of engineering structures. In this context, various probability distributions, such as the lognormal and Weibull distributions, have been proposed for fitting the data on fatigue life. (See Section 9.) It is clear, however, that such probability distributions (independently of how well they fit specific test results) only constitute representations of the dispersed life data and do not provide any direct relationship to the basic fatigue mechanisms.

A general mathematical approach to the analysis of the random fatigue process, including life prediction, is as follows. The development of fatigue in real materials subjected to realistic actions is, as a function of time, a certain nondecreasing stochastic process, say, $X(t)$, defined for $t \in [t_0, \infty)$. Fatigue failure occurs at a random time $t = T$, for which the process $X(t)$ crosses a fixed critical level x^*. Hence, the distribution of fatigue life is given as the probability distribution of the random variable $T = \inf\{t: X(t) \geq x^*\}$.

Therefore, stochastic modelling of fatigue (and of other deterioration processes) consists of three basic steps: (i) choosing an appropriate stochastic model-process for fatigue accumulation; (ii) determining the probabilistic properties of the model-process (e.g., its one-dimensional probability distribution or the distribution of T); (iii) relating the model-process to empirical data and parameter estimation. It is clear that the process $X(t)$, itself a proposed model-process, should be deduced from available empirical information about the fatigue process in real materials. Various possible approaches to the stochastic model building for fatigue will be presented in Chapters IV–VI.

* * *

The objective of this book is to present a unified approach to stochastic modelling of the fatigue phenomenon (especially, the fatigue crack growth process). The main approaches to construction of stochastic models of the fatigue process will be presented in an attempt to show their methodological consistency and potential usefulness in engineering practice. We hope that the analysis contained in this monograph will also have an inspiring influence on the development of new methodology for designing and performing fatigue experiments. In contrast to the traditional experiments in which the emphasis is on the cycle-dependent nature of fatigue accumulation, we mean the unconventional experiments in which fatigue is viewed as a specific nondecreasing random process, evolving in time and approaching (at some random time) its critical level. Such experiments would yield more insight into inherent randomness of the fatigue process and would provide reliable data for statistical inference (e.g., parameter estimation).

The organization of the book is as follows. In Chapter I, a synopsis is given of the basic existing knowledge on the fatigue phenomenon, including recent deterministic fatigue crack growth models. Attention is focused on features of the fatigue phenomenon that are of prime interest in stochastic modelling. Chapter II contains the mathematical prerequisites that are necessary in stochastic analysis of fatigue, including the basic probability distributions for fatigue life and the characterization of stochastic processes particularly relevant to fatigue. The next chapters (III–VII) are entirely directed toward a theory of random fatigue.

Since in many real situations, external loadings and applied stress conditions are randomly varying in time, Chapter III is devoted to a concise presentation of the common models of random fatigue loads, along with the most useful formulae for their statistical characteristics. Chapters IV–VI contain a systematic presentation of the basic ideas and methods in stochastic modelling of fatigue, with special attention being given to models of fatigue crack growth. Three classes of models are distinguished and analyzed: the so-called evolutionary probabilistic models based mainly on Markov process theory (Chapter IV), the cumulative jump models associated with compound counting processes (Chapter V), and the differential equation models resulting from randomization of the empirical crack growth equations (Chapter VI). In Chapter VII, we discuss the problems associated with comparisons of various stochastic models and of the associated probability distributions. Also, we show some possible engineering applications of the analysis presented in the earlier parts of the book.

Chapter I

Fatigue of Engineering Materials: Empirical Background

1. Introductory Remarks

In general, *fatigue* can be defined as a phenomenon that takes place in components and structures subjected to time-varying external loadings and that manifests itself in the deterioration of the material's ability to carry the intended loading.

The fatigue phenomenon today is deemed to originate in the local yield of the material or, in other words, in the sliding of atomic layers. This sliding is caused by a combination of dislocations and local stress concentrations. It is assumed that each slip, no matter how small, is connected with a small deterioration of the material structure. Under cyclic stress conditions, there is migration of dislocations that result in localized plastic deformations. Microscopic cracks are created that grow and join together to produce major cracks. *Nucleation* (or crack initiation) and *crack growth* are commonly regarded as basic causes of fatigue damage accumulation and ultimate fatigue failure. Nucleation and crack growth also constitute two principal phases in the fatigue damage process. (See Fig. 1.1.)

The fatigue crack nucleation process is not yet fully understood; however, several models have been proposed to explain the origin of fatigue cracks. (See [23], [36].) In these models, various dislocation mechanisms in the metallic structure play a dominant role. Usually, fatigue cracks occur on the free surface of the body at places of high stress concentration (e.g., inclusions, surface imperfections, grain boundaries, etc.). Depending upon the material properties and the applied loading, the nucleation phase can be

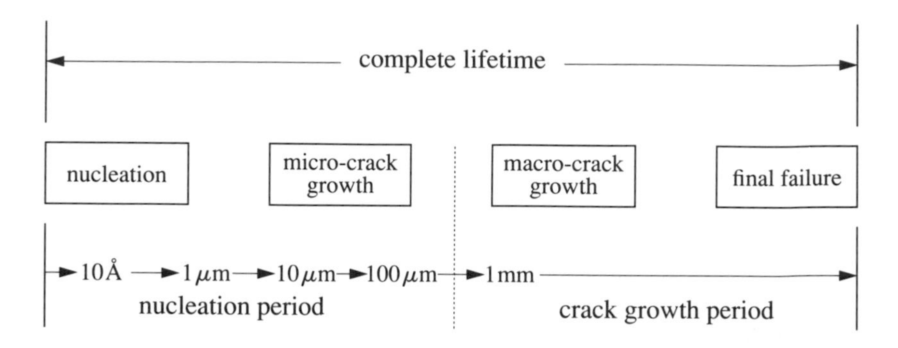

Figure 1.1. Fatigue life of a structural element.

of a different importance in estimating fatigue life. As indicated by experimental observations, at low stress levels (high-cycle fatigue), the crack initiation period may consume a significant percentage of the usable fatigue life, whereas at high stress amplitude (low-cycle fatigue), fatigue cracks start to develop in the early cycles. Also, in certain other structural details where defects are practically unavoidable due to the fabrication process, crack propagation may be considered to begin with the first load application. (See [9], [14].) In general, when the nucleation period cannot be neglected, the nucleation time should be estimated and included in the analysis of the total lifetime of a specimen. To this end, a weakest link model may be useful.

The question that arises when attempting to model the fatigue process is: When does a micro-crack become a macro-crack? As can be expected, no precise answer is possible; however, the following assertions give the necessary definition of the word *crack* as it is used in fatigue mechanics [38]:

- A crack is a macro-crack if it is large enough to be seen by the naked eye;
- a crack is a macro-crack if it has sufficient length to ensure that local conditions responsible for crack nucleation no longer affect crack growth;
- a crack is a macro-crack as soon as phenomenological fracture mechanics is applicable; i.e., when the stress intensity factor has a real meaning for describing crack growth.

On the macroscopic level, which is our main concern, the features of crack extension inferred from the experimental observations are:

- Cracks usually grow in a plane perpendicular to the largest principal stress (at least as long as the crack growth rate is not too high);
- crack growth is a repetitive and cumulative process that usually is concluded from the observation of striations on the fracture surface; each striation has been shown to represent the crack advance for one cycle of the load;
- in the process of crack extension, a plastic region is developed near the crack tip (due to high stress concentrations that exceed the yield strength of the material); this region is called the *plastic zone*;
- crack growth takes place mainly during the tensile (positive) part of the loading cycle; a compressive (negative) stress does not significantly affect crack growth;
- fatigue crack growth under constant-amplitude, cyclic loading can differ considerably from the growth under variable-amplitude loading (i.e., load interaction effects occur, e.g., retardation due to overloads);
- crack growth is affected by many factors of different nature; in general, crack growth is a function of the following variables:

  Stress range,

$$\Delta S = S_{\text{max}} - S_{\text{min}}, \tag{1.1}$$

  Stress ratio,

$$R = \frac{S_{\text{min}}}{S_{\text{max}}}, \tag{1.2}$$

  Stress complexity (variable amplitude, multiaxial, etc.),

  Mechanical properties (e.g., yield limit, elasticity, viscoelasticity),

  Metallurgical factors (e.g., alloy composition, heat treatment),

  Environmental factors (e.g., temperature, humidity, irradiation),

  Specimen geometry.

In what follows, we shall present a brief description of the basic methods in deterministic fatigue analysis, with emphasis being placed on information that will be used in the subsequent building of stochastic models.

2. Stress–Life (*S–N*) Diagrams

The first systematic and quantitative investigation of fatigue damage was provided by August Wöhler in 1858 and has resulted in the widely known *Wöhler curve*, or *S–N* curve (i.e., stress (*S*) versus number (*N*) of cycles to failure). This curve conveniently displays basic fatigue data in the elastic stress range (high-cycle fatigue) on a plot of cyclic stress (S) level versus number of cycles to failure (N). Constant amplitude S–N curves are usually plotted in semilog or log–log coordinate scales. (See Fig. 1.2, which shows a typical S–N diagram.) They carry important design information and are still used in engineering practice to estimate long lives of machine parts. To estimate the life data for high load levels where plasticity effects become important (i.e., for short lives or low-cycle fatigue), strain–life relationships are more appropriate. The transition between low- and high-cycle fatigue usually occurs between 10 and 10^5 cycles. Section 5 will discuss high-cycle fatigue in more detail.

Analytical representation of the S–N curves is commonly given in the form

$$NS^b = k \tag{1.3a}$$

or

$$N = kS^{-b}, \tag{1.3b}$$

where b and k are material parameters estimated from test data obtained using identical specimens.

Because of scatter in fatigue life data at any given stress level, it has been agreed that there is not just one S–N curve for a given material, but a family of S–N curves with probability of failure as the parameter. These curves are called S–N–P curves, or curves of constant probability of failure on a stress-versus-life plot.

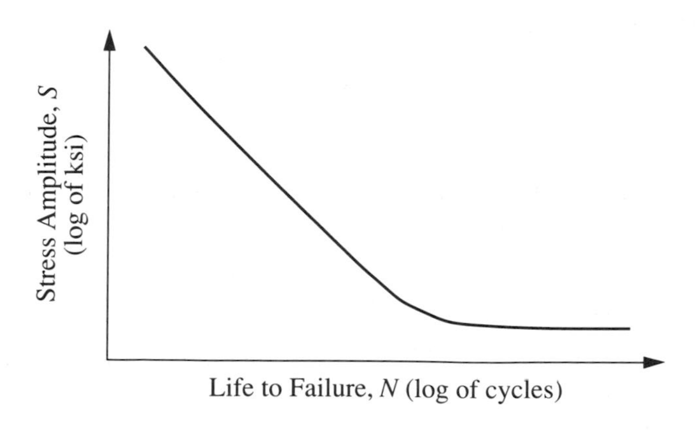

Figure 1.2. Typical S–N diagram for low-carbon steel.

Because of high uncertainty in the relationship between S and N, the parameters k and b in Eq. (1.3a) can be regarded as random variables (see [1] and [31]); in such a case, the statistical analysis leads to an expression for N in terms of the statistics of the dispersed data.

In structural steel design codes, a stress range threshold S_0 is often introduced. For stress levels below this threshold, no damage is assumed to occur (and an *infinite* life is assumed). The S–N relation (1.3b) is then

$$N = \begin{cases} kS^{-b}, & S > S_0 \\ \infty, & S \le S_0. \end{cases} \tag{1.4}$$

It should be realized that the S–N approach, though still widely used in design applications, does not deal with any of the physical phenomena within the material. For example, it does not separate the crack initiation from the propagation stage, and only the total life to fracture is considered.

3. Cumulative Damage Theories

3.1 Linear Damage Accumulation Rule

The stress amplitude experienced by a structural member may often vary during its service life. In such situations (i.e., under conditions of variable amplitude loading), the direct use of standard S–N curves is not possible. (These curves are presented for constant stress amplitude operation.) To estimate fatigue life in more general circumstances, Palmgren and Miner proposed that fatigue fracture is a result of a linear accumulation of *partial* fatigue damage.

The Palmgren–Miner rule asserts that the damage fraction, Δ_i, at any stress level S_i is linearly proportional to the ratio of n_i, the number of cycles of operation under this stress amplitude, to N_i, the total number of cycles that would produce a failure at that stress level; that is,

$$\Delta_i = \frac{n_i}{N_i}, \qquad n_i \le N_i. \tag{1.5}$$

If the stress amplitude is changed, a new *partial* damage is calculated for this new amplitude level. (The appropriate N_i is found from the S–N curve.) The total accumulated damage is then given by

$$D = \sum_i \Delta_i = \sum_i \frac{n_i}{N_i}. \tag{1.6}$$

It is assumed that failure occurs if $D \geq 1$, although typical values of D at failure range from 0.5 to 2.0. The life to failure can be estimated by summing the percentage of life used up at each stress level.

The weaknesses of the preceding hypothesis are obvious. Perhaps the most significant shortcoming of the Palmgren–Miner hypothesis is that it does not account for sequence effects; that is, it assumes that damage caused by a stress cycle is independent of where it occurs in the load history. It should also be noticed that the only material characteristic involved in this rule is the number of stress cycles to failure under constant amplitude loading. While the absolute validity of the Palmgren–Miner rule is in question, its comparative usefulness as a simple criterion for comparing different designs of a structure is still of considerable value.

3.2 Nonlinear Damage Hypotheses

To overcome the shortcomings of the Palmgren–Miner rule, a number of nonlinear damage hypotheses have been proposed. One of the first nonlinear damage rules was proposed by Marco and Starkey [33] and has the following analytical form:

$$\Delta_i = \left(\frac{n_i}{N_i}\right)^{x_s}, \tag{1.7}$$

where the exponent x_s is a function of the stress level and is assumed to fall in the range between zero and one (with the value increasing with stress level); Δ_i denotes a damage fraction associated with stress amplitude S_i. Failure, or 100 percent damage, is reached when the sum of Δ_i reaches a critical value. Analogous to the Palmgren–Miner rule, the values of N_i are the lives to failure corresponding to S_i on the S–N curve. A deficiency of this approach is that the family of stress curves on the damage plot must be developed experimentally for a given material. (See Bannantine et al. [3].) When the exponent x_s in Eq. (1.7) is independent of the stress condition (i.e., it is a constant value for all stress conditions), the damage specification (1.7) reduces to the so-called *modified Miner rule*.

A modification of Miner's rule referred to as Shanley's rule [43] is expressed as follows:

$$\Delta = cS^{kb}n, \tag{1.8}$$

where Δ is a damage resulting from n cycles applied at stress amplitude S. The quantity b is the shape parameter for the central portion of the S–N diagram, whereas c and k are material constants $(k > 1)$. The aforementioned expression for damage differs from the previous ones in that it is a function of the number of applied cycles rather that the cycle ratio.

A rule that is based on similar reasoning is known as the Corten–Dolan hypothesis [10]. It postulates (on the basis of data analysis) that damage due to n cycles of pure sinusoidal stress is given by

$$\Delta = rn^{\alpha}, \tag{1.9}$$

where r is a function of the stress level and is termed the coefficient of damage growth rate, and α is a constant. It is seen that the Corten–Dolan damage specification will coincide with Shanley's when $\alpha = 1$ and $r = cS^{kb}$.

Another approach to cumulative damage was proposed by Henry [18], who has postulated that fatigue damage may be defined as the ratio of the reduction of the fatigue limit (or endurance limit) to the original fatigue limit of virgin material; that is,

$$\Delta = \frac{F_0 - F}{F_0}, \tag{1.10}$$

where F_0 is the original fatigue limit and F is the fatigue limit after damage. After representing the S–N curve in the form $N = k_0/(S - F_0)$, where k_0 is a material constant, and after some additional reasoning (see Collins [9]), the Henry damage formula takes the form

$$\Delta = \frac{\frac{n}{N}}{1 + \left(\frac{F_0}{S - F_0}\right)(1 - \frac{n}{N})}, \tag{1.11}$$

where Δ is a damage fraction due to n cycles at an amplitude S; as defined previously, N denotes the number of cycles to failure. In the case of a sequence of different stress levels, the Henry rule is applied successively in the order of applied stress levels. In this sequential procedure, the value of F_0 must be updated after the application of each stress amplitude. As a result, a sequence of values for the fatigue limits, $F_0, F_1, F_2, \ldots$ is obtained that characterize a decrease of fatigue limit due to damage.

A number of other constructions of the nonlinear damage accumulation can be found in the literature. (See [9].) In general, they require material and shaping constants that have to be determined from a series of step tests. In more complicated situations (especially if one wants to take into account sequence effects), this requires a large amount of testing. For this reason, a simple linear Palmgren–Miner rule is still attractive for engineers.

3.3 Continuum Damage Mechanics

The damage accumulation rules were presented earlier in relation to fatigue due to loading at various amplitude levels. The concept of a cumulative damage, however, has a much wider meaning and is usually used to characterize globally all deterioration phenomena taking place in the material, both of the mechanical nature (fatigue, wear, creep, accumulation of plastic strain) and of the physical/chemical origin (e.g., corrosion, erosion, adsorption). Despite the diversity of these phenomena, it is useful to describe them jointly within a single model relating the rate of damage evolution with the acting loads and environmental conditions. Models of this type operate with a certain damage measure $D(t)$, which characterizes a damage state at time t. It is usually assumed that $D(t)$ is on the interval [0,1] and that it is a nondecreasing function of time. In general, the complex phenomena associated with cumulative damage cannot be described by a single scalar characteristic. So, in addition to scalar damage measures, vector and tensor measures may be necessary.

Assuming that the evolution of the function $D(t)$ at time t depends only on the values of $D(t)$ at some initial time instant t_o and on the external actions $Q(t)$, one can postulate the following general differential equation model for $D(t)$:

$$\frac{dD(t)}{dt} = f[D(t), Q(t)], \tag{1.12}$$

where the function $f(D,Q)$ is nonnegative and satisfies the appropriate conditions ensuring the existence and uniqueness of the solution of Eq. (1.12). Equation (1.12) can be regarded as a *kinetic equation* for damage evolution. (See [6].) This can be a scalar or vector equation, depending on the interpretation of $D(t)$ and $Q(t)$.

If the external actions (e.g., loading) change in a discrete way, the kinetic equation in (1.12) can be replaced by its finite difference analog; in such a case, a discrete variable (e.g., N, the number of cycles) plays the role of time.

When the right-hand side of Eq. (1.12) is independent of the damage measure D, then the solution of the equation with the initial condition $D(0) = 0$ represents a linear accumulation given by

$$D(t) = \int_0^t f[Q(\tau)]\, d\tau. \tag{1.13}$$

The time T at which the damage reaches its critical level can be determined from the condition $D(T) = 1$. Introducing the notation $t(Q) = 1/f(Q)$, we have the following equation for T:

$$\int_0^T \frac{d\tau}{t[Q(\tau)]} = 1. \tag{1.14}$$

The value $d\tau / t[Q(\tau)]$ characterizes a damage accumulated within the infinitesimal time interval $d\tau$. A discrete version of the preceding equation, when the external action $Q(t)$ is interpreted as multi-step level loading S_i, leads to the well-known linear cumulative damage rule of Palmgren–Miner. (See Bolotin [6] and Madsen et al. [31].)

The characterization of damage by the damage measure $D(t)$ is especially useful if we are interested in the deterioration of the material structure at early stages (e.g., during the initiation or nucleation phase of fatigue). Though damage during the initiation phase is related to dislocations, slip bands, micro-cracks, voids, and so on, these phenomena are not easily characterized quantitatively (e.g., by a specific observable and measurable quantity like crack length at the macroscopic-crack growth stage).

The exact description of the actual evolution of the damage pattern not only defies our capability, but it does not seem to be needed in view of the fact that the details of the damage process will, to a significant degree, differ from one experiment to another. Hence, instead of trying to reproduce the fine details of the micro-defect pattern, it appears reasonable to introduce a variable (treated as an internal variable of the material), or a set of variables reflecting the damage in an appropriately defined *smoothed* sense. This way of reasoning constitutes a base for the recently developed theory called *continuum damage mechanics*.

The first characterization of damage along this line was done by Kachanov [21] in 1958 (see also [22]), who introduced a scalar measure of damage $D(t)$ to characterize macroscopically an internal degradation of the material. On the basis of the hypothesis of isotropic damage (i.e., micro-cracks and voids are equally distributed in all directions), he defined $D(t)$ as follows:

$$D(t) = \frac{\mathcal{A}}{\mathcal{A}_0}, \qquad 0 \le D(t) \le 1, \tag{1.15}$$

where $\mathcal{A}_0$ is the cross-sectional area of the undamaged material, $\mathcal{A}$ is the cross-sectional area of the damaged material, and $\Psi = 1 - D = (\mathcal{A}_0 - \mathcal{A}) / \mathcal{A}_0$. A simple form of the kinetic equation for $D(t)$ given by Kachanov (in the case of uniaxial tension) is

$$\frac{dD}{dt} = \mathcal{A}\left[\frac{S}{1-D}\right]^n, \tag{1.16}$$

where $\mathcal{A} > 0$ and $n \geq 1$ are material constants. It is seen that Eq. (1.16) is a special case of the general differential equation model in Eq. (1.12). The idea of Kachanov has been developed by many authors in various important directions. (See Leckie [26], Lemaitre and Chaboche [28], Lemaitre [27] and Krajcianoviĉ [25].)

An important direction of the extensions to Kachanov's idea concerns attempts to account for the anisotropy of damage evolution. Most likely, the first characterization of damage by a second-order tensor is due to Vakulenko and Kachanov [53]. (See also Murakami and Ohno [35] and Sidoroff [46].) One of the central objectives in the papers cited is to incorporate damage into the general constitutive equations of the deformed solid body. In general, these equations can be represented as (see Grabacki [16])

$$\boldsymbol{\epsilon}(t) = \mathcal{F}\{\mathbf{S}, D\}, \tag{1.17}$$

where $\boldsymbol{\epsilon}$ is the strain tensor, $\mathbf{S}$ is the stress tensor, D is a damage measure (scalar or tensor), and $\mathcal{F}$ is an appropriate functional. The response of an undamaged material can be obtained for the special case when $D = 0$, i.e.,

$$\mathcal{F}\{\mathbf{S}, D = 0\} = \mathcal{F}_0\{\mathbf{S}\}. \tag{1.18}$$

The damaged response is assumed to be characterized by replacement of the stress tensor $\mathbf{S}$ in Eq. (1.18) by an *effective* stress tensor $\mathbf{S}^*$; that is,

$$\boldsymbol{\epsilon}(t) = \mathcal{F}\{\mathbf{S}, D\} = \mathcal{F}_0\{\mathbf{S}^*\}, \tag{1.19}$$

where

$$\mathbf{S}^* = \frac{\mathbf{S}}{1-D}, \quad D = D(t). \tag{1.20}$$

As can be seen, the aforementioned idea represents an uncoupled description of the deformation and damage. More generally, one can formulate coupled governing equations for stress, strain, and damage.

The preceding general continuum damage mechanics theories can be used in the case when damage is regarded as pure fatigue. Since it is rather difficult to determine the precise beginning of macro-crack growth, continuum damage theory is usually considered with respect to the entire fatigue process without distinguishing the nucleation and growth stages. If, however, one would like to use the continuum damage mechanics for characterizations of fatigue crack growth (after specifying its beginning), a possible definition of the damage could be, for instance, $D = a/a_{cr}$. Here, a_{cr} is the critical crack length, and a is the current crack length.

4. Elastic Fatigue Fracture

4.1 Stress Intensity Factor

As previously indicated, to estimate the fatigue life of a component, it is important to describe the growth of fatigue cracks. This requires the use of fracture mechanics principles to describe the crack extension and the stress distribution near the crack tip.

If the applied loading generates sufficiently low stress levels (not exceeding the limit of elasticity), then the stress distribution in a cracked element can be calculated by use of linear elasticity theory (i.e., assuming linear elastic behavior of the material element). The stress field near the crack tip can be divided into three basic types, each associated with a different mode of crack surface displacements. (See [3].) These three modes are: mode I – opening or tensile mode; mode II – sliding or shearing mode; mode III – tearing or antiplane mode. (See Fig. 1.3.) In the analysis of fatigue fracture, mode I is the most common and has received the greatest attention.

Using the methods of linear elasticity theory, the stresses at any point in the element can, in principle, be characterized. In general, they are complex

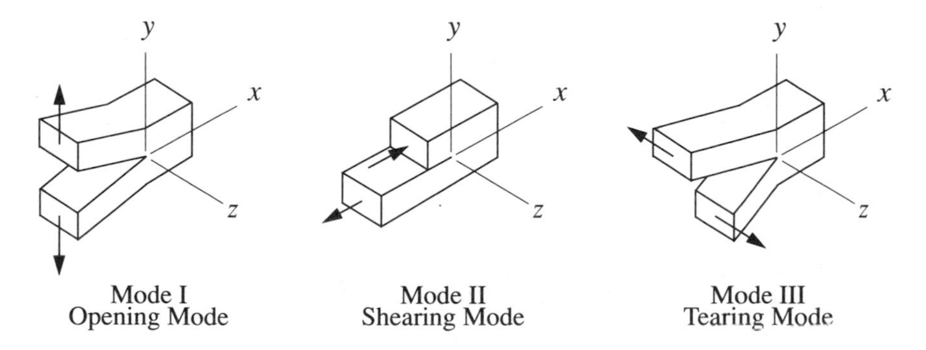

Figure 1.3. Three modes of crack extension.

and cannot be given in an explicit form. However, if we are interested in stresses only in the neighborhood of the crack tip (i.e., $r \ll a$; see Fig. 1.4), an asymptotic solution for the near stress field yields a formula for the stress components σ_{ij} that is relatively simple and is given by

$$\sigma_{ij} = \frac{K}{\sqrt{2\pi r}} f_{ij}(\theta) + \ldots, \tag{1.21}$$

where r and θ are cylindrical coordinates of a point with respect to the crack tip, and K is the *stress intensity factor*. This factor defines the magnitude of the local stresses around the crack tip. It depends on loading, crack size, crack shape, and geometry of the specimen. In general, the stress intensity factor has the form

$$K = BS\sqrt{\pi a}, \tag{1.22}$$

where a is the crack length, S represents the far-field stress resulting from the applied load, and $B = B(a)$ is a factor that accounts for the shape of the specimen and the crack geometry.

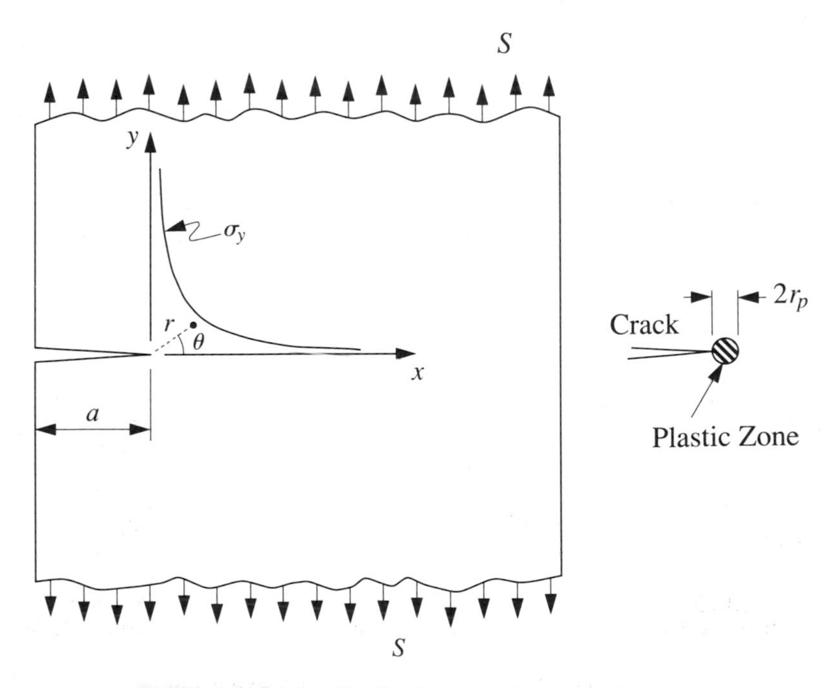

Figure 1.4. Stress distribution near the crack tip.

It is seen that the stress field has a singularity at $r = 0$ (i.e., $\sigma_{ij} \to \infty$ as $r \to 0$). Since infinite stresses cannot exist in a physical material, the elastic solution has to be modified to account for crack tip plasticity. Thus, a *plastic zone* around the crack tip is usually introduced. However, if the plastic zone is sufficiently small relative to the crack length and the specimen geometry, the stress field solution (1.21) can still be applied outside the plastic zone. The size and shape of the plastic zone can be estimated in terms of the stress intensity factor and yield stress. For the case of a purely plastic zone embedded in an elastic matrix, the radius r_p of the roughly circular plastic zone (under cyclic loading) is

$$\begin{aligned} r_p &= \frac{1}{2\pi}\left(\frac{K}{2\sigma_y}\right)^2, \quad \text{plane stress,} \\ r_p &= \frac{1}{6\pi}\left(\frac{K}{2\sigma_y}\right)^2, \quad \text{plane strain,} \end{aligned} \tag{1.23}$$

where σ_y is the yield stress. The plastic zone plays an important role in studying various phenomena associated with fatigue crack growth.

From Eq. (1.22), it follows that a cyclic variation of S will cause a similar cyclic variation of the stress intensity factor K. The basic quantities associated with K, such as $K_{\max}$, $K_{\min}$, ΔK, are given by

$$\begin{aligned} K_{\max} &= BS_{\max}\sqrt{\pi a}, \\ K_{\min} &= BS_{\min}\sqrt{\pi a}, \\ \Delta K &= K_{\max} - K_{\min}. \end{aligned} \tag{1.24}$$

Values of K for various loadings and geometrical configurations can be evaluated from elasticity theory utilizing both analytical and numerical calculations. The most common reference value of K is for a two-dimensional center crack of length $2a$ embedded in an infinite sheet and subjected to a uniform tensile stress S. In this case, $B = 1$ and

$$K = S\sqrt{\pi a} \approx 1.77S\sqrt{a}. \tag{1.25}$$

Units of K are MPa $\sqrt{\text{m}}$ and ksi $\sqrt{\text{in.}}$, where 1 MPa $\sqrt{\text{m}}$ = 0.91 ksi $\sqrt{\text{in}}$. For the single edge crack in a semi-infinite plate,

$$K = 1.12\ S\sqrt{\pi a} \approx 2S\sqrt{a}, \tag{1.26}$$

where the coefficient 1.12 in Eq. (1.26) is the free edge correction coefficient. Many other stress intensity factor expressions can be found in the literature. (See Fuchs and Stephens [14].)

The elliptical (embedded and surface) cracks are often taken as models of cracks in engineering components. The most common reference specimen in this context is the embedded elliptical crack in an infinite body subjected to uniform tension S at infinity. In this case, the stress intensity factor is given by [14]

$$K = \frac{S\sqrt{\pi a}}{\Phi}\left[\sin^2\beta + \left(\frac{a}{c}\right)^2\cos^2\beta\right]^{1/4}, \tag{1.27}$$

where β is the appropriate angle parameter, $2a$ is the minor diameter, and $2c$ is the major diameter; the form Φ is the complete elliptical integral of the second kind and depends on the ratio a/c.

Surface elliptical or semi-elliptical cracks can also play an important role in engineering. The stress intensity factor K associated with such cracks has been estimated mostly by numerical methods. (See Section 6.2.)

If a specimen is subjected to combined mechanical and thermal loading, the appropriate value of K can be obtained from solution of the thermoelastic problem. Similarly, if one is to account for the internal stresses (e.g., those generated in the material during fabrication process), the stress intensity factor should be supplemented by K_r — the stress intensity factor associated with residual (internal) stresses.

When the stress intensity factor reaches a critical value, K_c, unstable fracture occurs. This critical value of K is known as the *fracture toughness* of the material. This is an important material property, which characterizes a change from slow (stable) crack growth to rapid (unstable) growth. Thus, K_c can be regarded as a single-parameter fracture criterion; the fracture is predicted to occur if

$$BS\sqrt{\pi a} \geq K_c. \tag{1.28}$$

The fracture toughness K_c depends on the material, temperature, environment, and thickness of the specimen.

4.2 Empirical Fatigue Crack Growth Equations

Results of laboratory experiments and observations of in-service structures clearly indicate that fatigue damage (usually measured by the length of a dominant crack a) is affected by a variety of factors, e.g., the state of stress S, the material properties (denoted symbolically by C), temperature θ, and other environmental effects. In general, the fatigue crack growth can be characterized by the following nonlinear equation:

$$\frac{da}{dN} = F[a, S, C, \theta, \zeta], \tag{1.29}$$

where a, S, C, θ are the quantities indicated previously and ζ denotes symbolically all other parameters affecting fatigue crack growth (e.g., chemical properties, internal stresses, etc.); N denotes a number of cycles (or continuous time) corresponding to the crack length a. Unfortunately, there is not enough knowledge concerning the influence of each of the preceding variables on fatigue crack growth, and analytical forms of the general relations like Eq. (1.29) are lacking even for traditional fatigue problems.

The mathematical hypotheses concerning fatigue crack growth rate, deduced from experimental data, are commonly called the *fatigue crack growth equations*. The equations existing today in the literature have mostly been identified for constant-amplitude cyclic loading and have attempted to relate the crack growth rate da/dN with crack length, applied stress, and material parameters. The investigations of fatigue crack growth in elastic materials have shown that the most suitable quantity for characterizing the fatigue crack growth rate is the stress intensity factor K, or — more specifically — the stress intensity factor range $\Delta K = K_{max} - K_{min}$. Therefore, Eq. (1.29) takes the form

$$\frac{da}{dN} = F(\Delta K). \tag{1.30}$$

Among the equations of this form, the most widely accepted is the Paris–Erdogan equation,

$$\frac{da}{dN} = C(\Delta K)^m, \tag{1.31}$$

where C and m are regarded as material constants. However, it should be emphasized that though C and m are constant, they depend on many factors (e.g., material properties, environment, frequency, temperature, stress ratio), and their numerical values deviate from experiment to experiment. Moreover, the stress ratio $R = S_{min}/S_{max}$ was recognized to have a significant influence on fatigue crack growth. (See Bannantine et al. [3].) So, instead of Eq. (1.31), the equation for fatigue crack growth should be written as

$$\frac{da}{dN} = f(R, \Delta K). \tag{1.32}$$

In addition, experiments indicate that increasing the stress ratio R results in an increased crack growth rate; thus, Eq. (1.32) takes the form

$$\frac{da}{dN} = C\,g(R)\,(\Delta K)^{m}. \tag{1.33}$$

A number of laws have been proposed that fall into the general form of Eq. (1.33). One example is

$$\frac{da}{dN} = C\left(\frac{1}{1-R}\right)^{2}(\Delta K)^{3}. \tag{1.34}$$

Another proposed law is

$$\frac{da}{dN} = C\,(1+\alpha R)^{q}\,(\Delta K)^{q}, \tag{1.35}$$

where experiments with aluminum welds and some steels showed that $\alpha = 3/4$ and $3 \le q \le 7$.

An empirical equation that is regarded as one of the most notable and that accounts for the stress ratio effects is the Forman equation,

$$\frac{da}{dN} = \frac{C\,(\Delta K)^{m}}{(1-R)\,K_c - \Delta K}, \tag{1.36}$$

where K_c is a fracture toughness (a critical level of K corresponding to unstable fracture).

Considering the relation between da/dN and ΔK, there are two limitations on ΔK:

a) If ΔK becomes too large, a static failure will occur immediately, which implies that $K_{\max}$ has exceeded the fracture toughness, K_c, or equivalently, ΔK has exceeded $(1-R)\,K_c$;

b) the other limitation comes from the fact that cracks will not propagate if ΔK is too small; there is a threshold value ΔK_0, implying that crack growth requires $\Delta K > \Delta K_0$. (See Fig. 1.5.) Analytically, these limits can be written as

$$\begin{aligned} &\frac{da}{dN} \to 0 \quad \text{as} \quad \Delta K - \Delta K_0 \to 0, \\ &\frac{da}{dN} \to \infty \quad \text{as} \quad K_c - K_{\max} \to 0. \end{aligned} \tag{1.37}$$

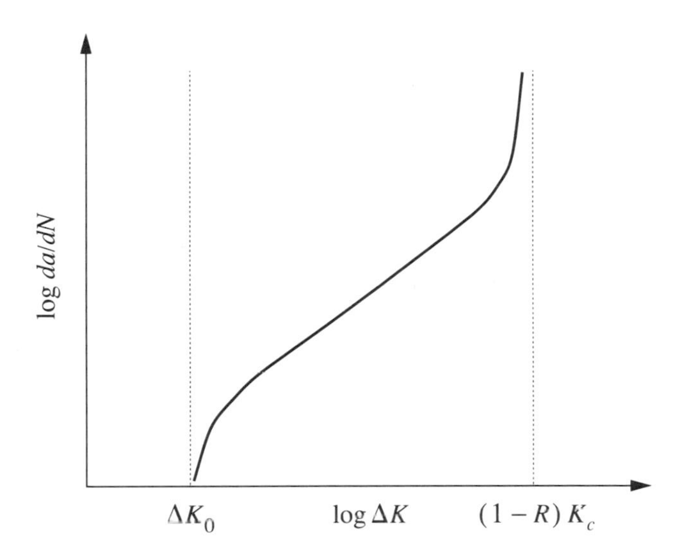

Figure 1.5. Limitations of ΔK.

To incorporate these limitations into crack growth equations, one can write the Forman equation as

$$\frac{da}{dN} = \frac{C\,(\Delta K)^{m}}{(1-R)\,(K_c - K_{max})} \tag{1.38a}$$

or

$$\frac{da}{dN} = \frac{C\,(\Delta K - \Delta K_0)^{m}}{(1-R)\,(K_c - K_{max})}\,. \tag{1.38b}$$

Some experiments have shown that ΔK_0 depends on the mean stress R, and the relation proposed is $\Delta K_0 = C_0\,(1-R)^{\beta}$, where C_0 is constant and $0.5 \le \beta \le 1.0$.

The Forman equation gives good results (compared with experiments) for the crack growth rate in steels and aluminium alloys. However, when the data of tests on viscoelastic materials were analyzed using this equation, no success was achieved.

Based on the outcome of an extensive test program covering a wide range of the stress intensity factors at various loading frequencies, the following relationship has been proposed by Radon et al. [37] for the polymer-like materials:

$$\frac{da}{dN} = \beta\lambda^m,$$
$$\lambda = (K_{max}^2 - K_{min}^2) = 2(\Delta K) K_{max}. \tag{1.39}$$

Let us return to the Paris–Erdogan equation (1.31), which is the simplest and the most often used. Based on a comparison with experimental data, the value of m is usually predicted to be in the interval between 2 and 4. C is the numerical factor dependent on loading conditions, the mean level of stress R, the material properties, and the characteristics of environment. It is clear that in the case of a homogeneous cyclic loading, the relationship between number of cycles N and time t is simple, namely, $N = \omega t$, and

$$\frac{da}{dt} = \frac{da}{dN}\frac{dN}{dt} = \omega\frac{da}{dN}, \tag{1.40}$$

where ω is a constant frequency. In the case of an infinite elastic sheet loaded at infinity by a tensile stress S, the Paris–Erdogan equation becomes

$$\begin{aligned}\frac{da}{dt} &= C\omega\pi^{1/2}(\Delta S)^m a^{m/2}\\ &= C_1(\Delta S)^m a^{m/2},\end{aligned} \tag{1.41}$$

where $C_1 = \omega\pi^{1/2}C$. The solution satisfying the initial condition $a(0) = a_0$ is

$$a(t) = \begin{cases} a_0 \exp\left[C_1(\Delta S)^2 t\right], & m = 2\\ \left[a_0^{(2-m)/2} + \dfrac{2-m}{2}C_1(\Delta S)^m t\right]^{2/(2-m)}, & m \neq 2.\end{cases} \tag{1.42}$$

It is seen that for $m > 2$, there exists a finite instant of time $t = \tau$ in which the solution explodes to infinity. From the solution just given, one easily obtains the time, or the number of load cycles, in which crack reaches a critical value a^*,

$$t(a^*) = \begin{cases} \dfrac{1}{C_1(\Delta S)^2}\ln\dfrac{a^*}{a_0}, & m = 2\\ \dfrac{2}{2-m}\dfrac{1}{C_1(\Delta S)^n}\left(a^{*(2-m)/2} - a_0^{(2-m)/2}\right), & m \neq 2.\end{cases} \tag{1.43}$$

It is worth noting that at failure

$$t(a^*)\,(\Delta S)^m = \text{constant}. \tag{1.44}$$

The solution (1.42) is completely determined by the five parameters, a_0, $C_1(C, \omega)$, n, ΔS.

To make the expression for fatigue life in Eq. (1.43) entirely explicit, the critical crack size a^* must be estimated. This may be done using Eq. (1.22) and the definition of the fracture toughness K_c; namely,

$$a^* = \frac{1}{\pi}\left[\frac{K_c}{BS_{\max}}\right]^2. \tag{1.45}$$

If the correction factor B varies with a, an iterative procedure is required to obtain the value of a^*.

Remark: The formulae in this section show that crack growth is clearly a nonlinear phenomenon, independent of the fact that the basic governing quantity, K, is obtained from linear elasticity theory.

4.3 Crack Growth under Variable-Amplitude Loading

The traditional fatigue crack growth equations — including these presented earlier — are based on fixed stress level fatigue experiments (i.e., under constant-amplitude, homogeneous, cyclic loading). They do not take into account so-called *interaction effects* due to irregularity of loading, including random loading. In contrast to the constant-amplitude loading, under variable-amplitude loading, the increment of fatigue crack growth depends, in general, not only on the present crack size and applied load, but also on the preceding load history. Load interaction or sequence effects have significant influence on the fatigue crack growth rate and, consequently, fatigue life.

Various types of variable-amplitude loading occur in reality (or are generated in laboratories to increase the understanding of the crack growth mechanisms). These loadings can be classified as follows: *step loading, flight-simulation load* (sequence of loads in flights), *programmed block loads, cyclic load with overloads,* and *random loading*.

One of the important interaction effects (recognized in the early 1960s) is the *retardation* in fatigue crack growth following a sufficiently large tensile overload. (See Fig. 1.6.) Crack retardation remains in effect for some period after the overloading. The number of cycles in the retarded growth has been shown to be related to the plastic zone size developed due to the overload. The larger the plastic zone generated by the overload, the longer

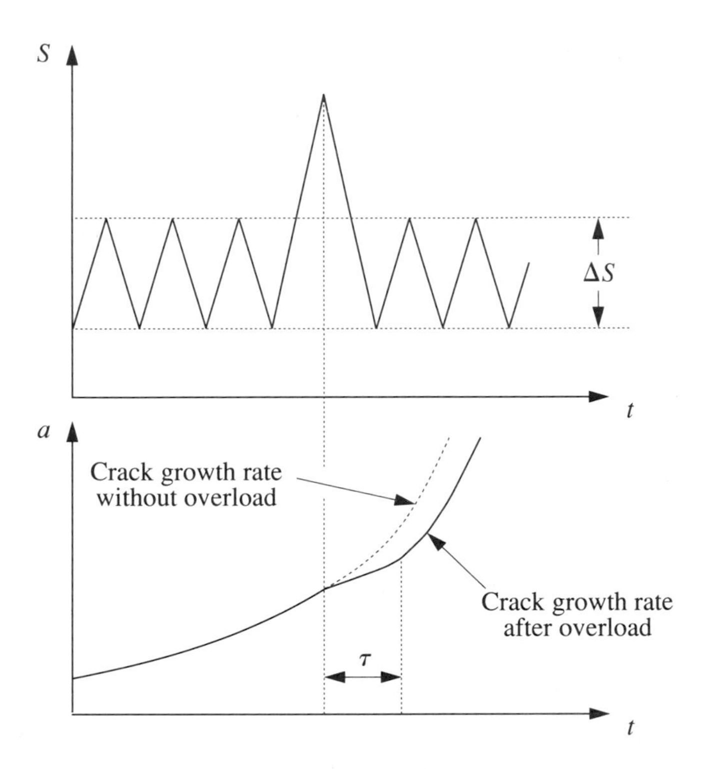

Figure 1.6. Crack growth retardation due to overload.

the crack growth retardation remains in effect. In general, longer retardation occurs with increasing magnitude of the overload or with repeating the overload during the crack growth period (multiple overloads). Application of blocks of overloads (instead of a single one) also extends the retardation period. Most often, crack growth retardation does not occur immediately after an overload; usually, a *delayed retardation* occurs.

If the overload is very large, *crack arrest* can occur and the growth of the fatigue crack stops for some time. Negative overloads (compressive stress conditions) have relatively small effect on crack growth; but if a compressive load is large enough to cause yielding at the closed crack tip, the subsequent crack growth will be *accelerated*. (See Stephens et al. [51].) When a tensile overload immediately follows a compressive overload, the effect of retardation is greatly diminished. Observations indicate that the acceleration effect (due to negative overloads) is not as great as the retardation effect (due to positive overloads); as a result, it attracts less attention. (See Trebules et al. [52].)

An important question is: How does one predict fatigue crack growth for variable amplitude loading? There is no general standardized method, but a number of procedures have been put forward that have some predictive ability.

Since the stress intensity factor range ΔK plays a key role in the case of constant amplitude loading, it is usually believed that this quantity should also be a factor in hypotheses concerning fatigue crack growth under variable amplitude loading. Several of the prediction methods will be outlined in the following paragraphs. (See Schijve [41].)

4.3.1 Cycle-by-Cycle (Noninteractive) Prediction

The crack length is represented as

$$a_N = a_0 + \sum_{i=1}^{N} \Delta a_i, \tag{1.46}$$

where a_0 is the initial crack length (for $N = 0$), and $\Delta a_i = da/dN$ for a given ΔK and R in cycle i as predicted under constant-amplitude loading. The deficiency of such a scheme is that the Δa_i are independent of the history of the preceding crack growth.

4.3.2 Crack Closure Model of Elber

This is an empirically based model that uses an *effective stress range* concept. The idea here is that the crack will propagate only when the stress is greater than the stress that causes the crack faces to fully separate or open. The effective stress range is then defined as

$$\Delta S_{eff} = S_{\max} - S_{op}, \tag{1.47}$$

where S_{op} is the opening stress that is determined experimentally. If we designate

$$c_f = \frac{S_{op}}{S_{\max}} = \frac{K_{op}}{K_{\max}}, \tag{1.48}$$

then

$$\Delta S_{eff} = S_{\max}(1 - c_f), \tag{1.49}$$

and, if the Paris–Erdogan equation is used, we have

$$\frac{da}{dN} = C\,[\Delta K_{eff}]^m = C\,[K_{\max} - K_{op}]^m$$
$$= C\,[\,(S_{\max} - S_{op})\,B\sqrt{\pi a}\,]^m \tag{1.50}$$
$$= C\,[S_{\max}\,(1 - c_f)\,B\sqrt{\pi a}\,]^m.$$

If S_{op} is defined as a function of the previous load history, then the preceding equation would predict crack growth interaction effects. This model is often used to predict retarded crack growth caused by overloads.

4.3.3 Wheeler Model

To predict retarded crack growth due to an overload, Wheeler has assumed that retardation occurs as long as the monotonic plastic zone size is smaller than the overload plastic zone. The retarded crack growth rate is related to the growth rate associated with the constant-amplitude loading through

$$\left(\frac{da}{dN}\right)_{OL} = C_p\left(\frac{da}{dN}\right)_{CA}, \tag{1.51}$$

where the retardation factor, $C_p \leq 1$, is a function of the ratio of the current plastic zone size to the plastic zone size created by the overload; explicitly,

$$C_p = \left[\frac{r_p}{(a_* + r_{po}) - a}\right]^p = \begin{cases} \left(\dfrac{r_p}{s-a}\right)^p, & r_p < s - a \\ 1, & r_p \geq s - a, \end{cases} \tag{1.52}$$

where r_p is the size of the plastic zone at the crack length a and corresponding to constant-amplitude loading, r_{po} is the size of the plastic zone caused by the overload occurring at the crack length a_*, $s = a_0 + r_{po}$, and p is an empirical constant. (See Fig. 1.7.)

4.3.4 Concept of Equivalent K

The equivalent K concept was proposed mainly for randomly irregular loading. The basic assumption is that an equivalent ΔK can be found, which under constant-amplitude (CA) loading will give the same crack rate as the random loading; that is,

$$\left\{\frac{da}{dN} = f(\Delta K_{eq})_{\substack{random\\ loading}}\right\} = \left\{\frac{da}{dN} = f(\Delta K)_{\substack{const\\ ampl\\ loading}}\right\}. \tag{1.53}$$

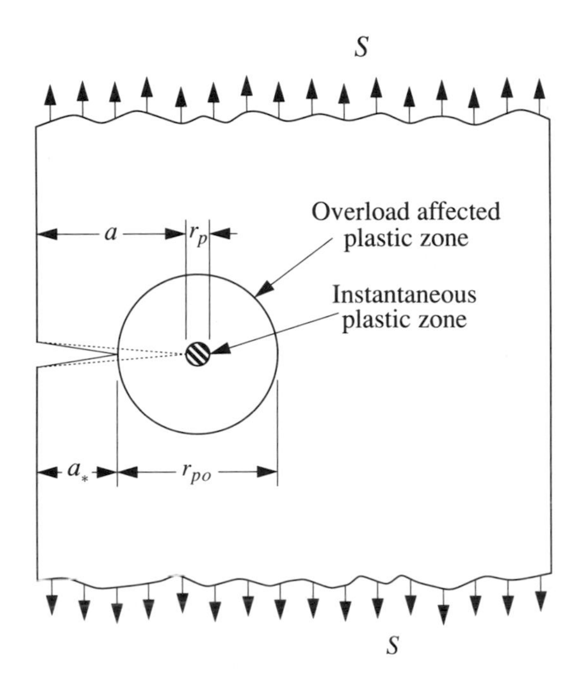

Figure 1.7. Crack tip plastic zone in Wheeler model.

Most often, the root mean square (rms) value of random loading was proposed for this purpose, i.e., $\Delta K_{eq} = \Delta K_{rms}$, and a relationship to constant-amplitude data is abandoned. In this case, the fatigue crack growth equation is postulated in the form

$$\frac{da}{dN} = f(\Delta K_{rms}), \qquad \Delta K_{rms} = S_{rms} B \sqrt{\pi a}. \tag{1.54}$$

There are also some other particular proposals to describe crack growth in complicated loading situations (e.g., Willenborg model, residual force model, etc.). It seems that none of these empirical models is accepted as being superior to the others. It is important to formulate a more uniform approach to modelling and analysis of fatigue crack growth. Most likely, this aim can be realized with use of probabilistic reasoning.

5. Plasticity Effects

5.1 Strain–Life Curves

In the previous sections, attention has been focused on situations where the loading amplitude is relatively low (i.e., stresses and strains are related according to linear elasticity theory). In these cases, a long life or high number of cycles to failure is exhibited (*high-cycle fatigue*).

Although the usual objective of engineering design is to provide long life, there are many circumstances in which high stress concentrations occur and a significant amount of plastic strain is generated. In such cases, short lives are exhibited when these high-level stresses are repeatedly applied. (This type of fatigue is commonly called *low-cycle fatigue*.) In the regions where plastic deformation is significant, the fatigue life can be more adequately described as a function of the strain amplitude (not the stress amplitude). Such strain–life approaches have gained wide acceptance as useful methods for evaluating the fatigue life of notched components.

Analysis of experimental data has led to the empirical relationships between plastic strain range $\Delta\epsilon_p$ and the number of cycles to failure N_f. The most well-known is the relation independently proposed by Manson and Coffin in 1954,

$$\frac{\Delta\epsilon_p}{2} = \epsilon_f'(2N_f)^c, \tag{1.55}$$

where $\Delta\epsilon_p/2$ is the plastic strain amplitude, ϵ_f' is the fatigue ductility coefficient, and c is the fatigue ductility exponent. (ϵ_f' and c are empirical constants characterizing fatigue properties of the material.)

The *total strain* is the sum of the elastic and plastic strains, that is,

$$\frac{\Delta\epsilon}{2} = \frac{\Delta\epsilon_e}{2} + \frac{\Delta\epsilon_p}{2}. \tag{1.56}$$

Taking into account that

$$\frac{\Delta\epsilon_e}{2} = \frac{\Delta\sigma}{2E}, \tag{1.57}$$

and the following representation of the *S–N* curve [3],

$$\frac{\Delta\sigma}{2} = \sigma_f'(2N_f)^b, \tag{1.58}$$

we have

$$\frac{\Delta \epsilon}{2} = \frac{\sigma_f'}{E}(2N_f)^{b} + \epsilon_f'(2N_f)^{c}. \tag{1.59}$$

Here, E is Young's modulus, σ_f' is the fatigue strength coefficient, and b is the fatigue strength exponent. The preceding formula is known as the *strain–life relation* and serves as a tool of estimating the fatigue life in the presence of plastic deformation. It has the advantage of making it possible to describe both low-cycle and high-cycle fatigue within one mathematical scheme. Practical application of Eq. (1.59) — widely described in the literature (see [3], [14]) — requires estimation of parameters σ_f', b, ϵ_f', and c from appropriate fatigue experiments.

Like the stress–life diagrams in the elastic case, the strain–life relation in Eq. (1.59) does not display mechanisms of elastic–plastic fatigue accumulation. To gain a deeper insight into the fatigue process when crack tip plasticity is large enough to violate the applicability of linear elasticity, empirical relations were proposed in the 1960s that relate the fatigue crack growth rate with the plastic strain range $\Delta\epsilon_p$. The general form of these equations is

$$\frac{da}{dN} = F(\Delta\epsilon_p). \tag{1.60}$$

Other proposals relate crack extension with so-called crack opening displacement (see Rolfe and Barsom [40]) and with the J-integral.

5.2 *J*-Integral and Fatigue Crack Growth

The path-independent integral, proposed by Rice in 1968, describes the intensity of the stress–strain field at the tip of a crack under elastic–plastic loading conditions. This integral is defined for either elastic or plastic behavior (within deformation theory of plasticity) as follows:

$$J = \int_C W dy - \mathbf{T}\left(\frac{\partial \mathbf{U}}{\partial x}\right) ds, \tag{1.61}$$

where C is a closed contour (followed counterclockwise) around the crack tip as shown on Fig. 1.8; $\mathbf{T}$ is the traction vector perpendicular to C, $T_i = \sigma_{ij}n_j$, $\mathbf{U}$ is the displacement vector, *ds* is the differential arc length along C, and W is the strain energy density, i.e.,

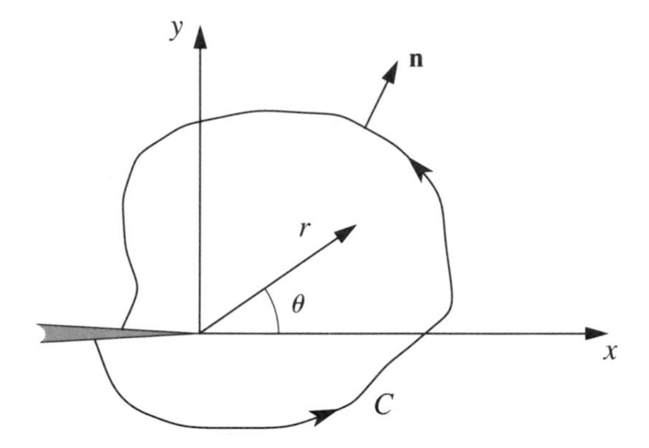

Figure 1.8. Line integral contour around the crack tip.

$$W(\epsilon_{ij}) = \int_0^{\epsilon} \sigma_{ij} d\epsilon_{ij}. \tag{1.62}$$

For the linear elastic case, the J-integral is identical to G, the energy release per unit crack extension, i.e. (for plane-strain),

$$J_e = G_e = \frac{(1-\nu^2)K^2}{E}, \tag{1.63}$$

where ν and E are Poisson's ratio and Young's modulus, respectively.

The usage of the J-integral for predicting fatigue crack growth in the presence of plastic deformations is not straightforward. The reason is that the J-integral is defined (and is path-independent) for elastic (linear or non-linear) materials as well as for plastic behavior within deformation theory of plasticity (small strains). In general, plasticity can only be described adequately by an incremental (flow) theory of plasticity; but momentary stresses and strains are then no longer uniquely related (as in elasticity theory or in deformation plasticity). Therefore, the strain energy density W, according to the definition in Eq. (1.62), is no longer a function of momentary strains, but becomes dependent on the loading history. Therefore, the stresses cannot be obtained from W by differentiation. As a result, path independence of J cannot be proven.

Thus, one might expect the J-integral to be applicable in the case when the results of incremental plasticity theory are close to the results of deformation plasticity theory, and Eq. (1.62) is valid in a practical sense. Such a closeness of responses can be gained for monotonically increasing loads (i.e., without unloading) that are proportional to a single parameter.

There are two basic criteria for the validity of J-controlled crack growth. (See Hutchinson and Paris [19] and Dowling and Begley [13].) The first one is that the regions of elastic unloading and of distinctly non-proportional plastic loading (which are of the order of Δa) are small, i.e.,

$$\frac{\Delta a}{R} \ll 1, \tag{1.64}$$

where R is the size of the region in which the nearly proportional (described by the J-integral) load dominates. The second condition is that J should increase sufficiently rapidly with crack extension (to make the region of nonproportional loading small); this condition can be stated as [13]

$$\omega = \frac{b}{J}\frac{dJ}{da} \gg 1, \tag{1.65}$$

where b denotes the uncracked ligament.

The aforementioned restrictions for J-controlled crack growth expressed in general terms constitute a frame for possible applications. The actual limits depend on the type of structure and on the loading. Numerical analysis shows (see Shih and German [44]) that for bending loads, the conditions are $\omega > 25$, $\Delta a < 0.06b$, whereas for tensile loads, $\omega > 80$, $\Delta a < 0.06b$.

If the conditions stated previously are satisfied and the J–R curve (the plot of J versus Δa) may be regarded as a material response characteristic, then the crack growth may be predicted with the use of the J values. This has been confirmed by numerous studies. For example, Dowling and Begley [13] showed that there exists a good correlation between the J-integral range, ΔJ, and the fatigue crack growth rate for tests performed in the elastic–plastic regime. Similar observations have more recently been reported by Jablonski [20] on the basis of fatigue crack growth tests for 304 stainless steel. The results of his tests are represented by the equation

$$\frac{da}{dN} = 1.81 \times 10^{-9} \, (\Delta J)^{1.46}. \tag{1.66}$$

In general, the governing equation can be written in a form similar to the Paris–Erdogan equation as

$$\frac{da}{dN} = C \, (\Delta J)^{m}. \tag{1.67}$$

Since $\Delta J = \Delta J_e + \Delta J_p$, Eq. (1.67) takes the form,

$$\frac{da}{dN} = C\left[\frac{(\Delta K)^2(1-\nu^2)}{E} + \Delta J_p\right]^m. \tag{1.68}$$

To use Eq. (1.67) for predicting fatigue crack growth in specific situations, the value of ΔJ_p has to be determined. In some cases, it can be done analytically (see Shih and Hutchinson [45]), but in most practical situations (e.g., complex specimen geometry), the appropriate experimental techniques have been developed. (See Rice et al. [39] and Clarke and Landes [8].)

In recent years, a number of studies have been performed to improve predictive capabilities of elastic–plastic fracture mechanics. In addition to J-integral techniques, some other parameters have been introduced and examined through a combined experimental–numerical approach; one of them is the ΔT^*-integral introduced by Atluri [2]. (See also Brust et al. [7].)

6. Other Problems

6.1 Environmentally Assisted Crack Growth

There are many environmental factors that have significant influence on the fatigue process in engineering structures (e.g., corrosion, temperature, irradiation, etc.). One of the most important damaging phenomena occurring in structures interacting with air and water is corrosion. *Corrosion fatigue* refers to the joint deterioration process that occurs during interaction of corrosive environment and fluctuating stresses.

Experiments show that, in general, many aluminum, titanium, and steel alloys are adversely affected by the presence of corrosive environment. In these studies, test frequency, load ratio, load profile, and temperature have been recognized as major variables affecting fatigue crack growth in specimens subjected to aggressive environment. It has also been found that environmental acceleration of fatigue crack growth usually does not occur below a certain threshold value (K_{scc}) of the stress intensity K.

Due to the great expense and time required to obtain crack growth rates, there is not enough realistic crack growth data for corrosion fatigue. This lack of reliable data makes it difficult to build theoretical models of the process. Among the existing proposals, we wish to mention here the following empirical equation (see Rolfe and Barsom [40]):

$$\frac{da}{dN} = \varphi(t)\,(\Delta K)^m, \tag{1.69}$$

where $\varphi(t)$ characterizes the effects of environment and m is taken to be equal to 2. Another model proposed to account for the effects of environment on corrosion fatigue crack growth has the form (Wei and Landes — see Crooker and Leis [11])

$$\left(\frac{da}{dN}\right)_t = \left(\frac{da}{dN}\right)_r + \left(\frac{da}{dN}\right)_{ssc} + \left(\frac{da}{dN}\right)_{cf}, \tag{1.70}$$

where

$\left(\frac{da}{dN}\right)_t$ = total corrosion fatigue crack growth rate,

$\left(\frac{da}{dN}\right)_r$ = fatigue crack growth rate in an inert, or reference, environment (contribution of *pure* mechanical fatigue),

$\left(\frac{da}{dN}\right)_{scc}$ = stress-corrosion crack growth rate (at K levels above K_{scc}),

$\left(\frac{da}{dN}\right)_{cf}$ = cycle-dependent contribution resulting from synergistic interaction of fatigue and environmental attack (assumed to be proportional to the amount of hydrogen produced by the surface reactions during each cycle, which in turn is proportional to the effective crack area).

It is clear that corrosion itself is a complex physicochemical phenomenon resulting from interaction of metals with the environment. In the presence of time-varying (fatiguing) external loads, the electrochemical and physical phenomena interact with mechanical processes and result in accelerated crack growth. However, an adequate quantitative description of this complicated process creates serious difficulties. A possible phenomenological approach to formulate a general description of the corrosion fatigue process could be based on the idea of continuum damage mechanics (presented in Section 3.3). An attempt in this direction has recently been made by Bolotin [5]. (See also [6].) The idea is as follows.

Consider an edge crack at the surface of a body with a length (depth) a and a radius of curvature at the crack tip of ρ. Assume that the body is subjected to plane strain conditions with nominal stress S and to the environmental (corrosion) action characterized by quantity C (e.g., concentration of active agents on the surface). The growth of a crack is controlled by the accumulation of damage in the vicinity of the crack tip. Let ψ_s characterize *purely* mechanical fatigue damage and ψ_c the corrosion damage. Both measures assume values from the interval [0,1] and they depend on the load history. Processes $\psi_s(t)$ and $\psi_c(t)$ should be governed by the following equations:

$$\frac{d\psi_s}{dt} = F_s(S, C, \rho, \psi_s, \psi_c),$$
$$\frac{d\psi_c}{dt} = F_c(S, C, \rho, \psi_s, \psi_c), \tag{1.71}$$

where F_s and F_c are the appropriate functions of given processes $S(t)$ and $C(t)$ and of the required functions $\rho(t)$, $\psi_s(t)$, $\psi_c(t)$. Eqs. (1.71) should be supplemented by the equation for $\rho(t)$; such an equation — relating $\rho(t)$ with damage measures ψ_s and ψ_c — has been constructed in [6]. Also, some specific forms of governing functions F_s and F_c have been given. Correlation of the model described to the experimental test data seems to be an open question.

6.2 Surface Crack Growth

An important class of problems concerning fatigue life prediction of engineering structures is associated with analysis of surface cracks. The shape of the surface cracks varies during the fatigue process; hence, the estimation of this variability is of interest.

To make the analysis of the shape change possible, the surface cracks are usually assumed to be semi-elliptical, with the depth a and half-length c. (See Fig. 1.9.) It is also assumed that the crack growth rate at any point along the crack front obeys the Paris equation. This means that, at each specific point along the crack front, the increment in crack size dr_φ during the load cycle dN is related to the range of the stress intensity factor ΔK as follows:

$$\frac{dr_\varphi}{dN} = C(r, \varphi)\,[\Delta K(r, \varphi)]^m, \tag{1.72}$$

where $C(r, \varphi)$ and m are empirical constants (for that specific point along the crack front).

An additional simplifying assumption is that the fatigue crack (being initially semi-elliptical) remains semi-elliptical during the process of growth; so, the crack depth (a) and the crack length ($2c$) are sufficient parameters for the description of the crack front. As a consequence, general equation (1.72) is replaced by the pair of two coupled differential equations,

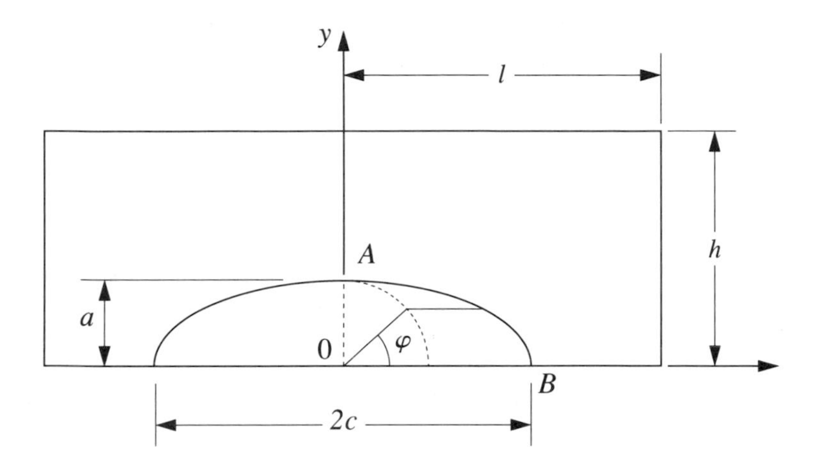

Figure 1.9. Surface semi-elliptical crack.

$$\frac{da}{dN} = C_A (\Delta K_A)^m,$$
$$\frac{dc}{dN} = C_B (\Delta K_B)^m. \tag{1.73}$$

The subscripts A and B refer to the deepest point and an end point of the crack at the surface, respectively. The constants C_A and C_B may differ due to the generally triaxial stress field. It is often assumed that $C_A = 1.1^m C_B$. (See Shang-Xian [42].)

To make Eqs. (1.73) explicit, the appropriate expression for the stress intensity factor for a surface semi-elliptical crack must be known. Because of the complexity of the surface crack problems, no exact analytical solution is available. Various numerical techniques have been used to estimate K and the results compared with experimental data. (See Mahmoud and Hosseini [32].) The solution that seems to be the most reliable is that of Raju and Newman [38], obtained by use of three-dimensional finite elements. This is an approximate expression for the stress intensity factor K for a surface crack in a finite plate subjected to a remote uniform tensile stress S_t and a remote bending stress S_b. The general form of this expression is

$$K(\varphi) = (S_t + HS_b)\sqrt{\frac{\pi a}{Q}}F\left(\frac{a}{h}, \frac{a}{c}, \frac{c}{l}, \varphi\right), \tag{1.74}$$

where φ is the angle that defines the position of the point at the crack contour, h is the thickness of the plate, and l is the half width of the panel; Q is the shape factor, whereas F and H define the boundary-correction factors. Equation (1.74) holds for $0 < a/c \leq 1.0$, $0 \leq a/h \leq 1.0$, $c/l < 0.5$, and $0 \leq \varphi \leq \pi$. Explicit expressions for Q, H, and F are provided in [38].

6.3 Multiaxial Fatigue

In many applications, structural components are subjected to complicated states of stress and strain (e.g., crankshafts and rear axles are usually subjected to combined bending and torsion). Complex, triaxial stress–strain states occur often in notches or joint connections. Fatigue under such conditions — termed *multiaxial fatigue* — constitutes an important problem in engineering practice.

Though in the past two decades the problem of multiaxial fatigue has attracted the attention of many researchers, the state of knowledge regarding this type of fatigue is still not satisfactory. This is mainly due to the difficulties involved in obtaining reliable experimental multiaxial fatigue data. Analysis of multiaxial fatigue requires an understanding of the triaxial state of stress and strain, along with all the phenomena that accompany the nucleation and growth of cracks.

Due to the complexity of the process, early studies of multiaxial fatigue were based on extensions of static yield theories to fatigue under combined stresses (e.g., the Tresca and von Mises criteria). To apply the data from uniaxial tests to multiaxial situations, authors have proposed some rules that reduce complex multiaxial loading to an *equivalent* uniaxial loading. (See Sines method, and Langer method [3].)

As far as fatigue cracks (nucleation and growth) are concerned, it has been recognized that the cracks initiate on planes with critical combinations of shear and normal stresses. The stresses on these planes also affect the fatigue crack growth.

Possible models for fatigue crack growth should, in general, account for all three loading modes (tensile, shearing, tearing). To extend the studies of fatigue crack growth from uniaxial stress states to the mixed mode situations, Sih [47] postulated that the strain energy density at the crack tip should be considered and that the rate of crack growth is a function of the strain energy factor range. As a result, the equation for the crack growth has the form

$$\frac{da}{dN} = f(\Delta W), \tag{1.75}$$

where ΔW is the strain energy density factor range given by $\Delta W = W_{max} - W_{min}$; W_{max} and W_{min} are the strain energy density factors associated with maximum and minimum stresses, respectively. The strain energy density W is defined as

$$W = d_{11}K_{I}^{2} + 2d_{12}K_{I}K_{II} + d_{22}K_{II}^{2}, \tag{1.76}$$

where K_I and K_{II} are the mode I and mode II stress intensity factors, respectively; d_{ij} contains the elastic constants (shear modulus and Poisson ratio) and varies with the polar angle θ measured around the crack tip.

Another possible approach to quantifying the crack growth rate under combined mode loading conditions stems from the observation that changes in crack rate under different cyclic stresses are a consequence of changes in crack tip plasticity. So, the growth equation is written in the form (see Sih and Theorcaris [48])

$$\frac{da}{dN} = \alpha r_p(\theta^*), \tag{1.77}$$

where $r_p(\theta^*)$ characterizes the maximum extent of the crack tip plasticity zone (θ^* is chosen to give the maximum value of r_p); the expression for $r_p(\theta)$ is given by

$$\begin{aligned} r_p(\theta) = \frac{1}{2\pi\sigma_y^2}\Bigg\{ & \Delta K_I^2 \cos^2\left(\frac{\theta}{2}\right)\left[(1-2\nu)^2 + 3\sin^2(\theta/2)\right] \\ & + \Delta K_I \Delta K_{II} \sin\theta\left[3\cos\theta - (1-2\nu)^2\right] \\ & + \Delta K_{II}^2\left[3 + \sin^2\left(\frac{\theta}{2}\right)\left((1-2\nu)^2 - g\cos^2\left(\frac{\theta}{2}\right)\right)\right]\Bigg\}. \end{aligned} \tag{1.78}$$

Hence, the rate of mixed mode crack growth is the maximum plastic zone size modified by a fracture ductility coefficient α. More information on multiaxial fatigue can be found in Miller and Brown [34].

Closing this section, it is worth noting that multiaxial fatigue damage can also be considered within the framework of continuum damage mechanics. In this context, the paper by Sorensen [50] may be of some interest. (The author describes a general linear isotropic cumulative damage process, including multiaxial stressing; the incremental fatigue damage is defined in terms of a suitable invariant function of a stress tensor.) Also, the analysis of Bolotin [6] is performed in the spirit of continuum damage mechanics.

7. Scatter in Fatigue Data

Experimental data obtained from testing specimens under various loading conditions constitute the main source of information about fatigue of engineering materials. However, these data, regardless of how carefully they are generated, show significant random scatter that may conceal their informational content. Scatter in fatigue test data is therefore a very significant issue in the analysis of the fatigue phenomenon and in prediction of fatigue reliability of engineering structures. (See [17], [29].)

Three sources of variability in experimentally obtained fatigue data are commonly regarded as the most decisive: namely, (i) the difference in material behavior among identically prepared specimens (due to difference in stress concentration at grain boundaries, effects of thermal processing, etc.); (ii) uncertainty in the fatigue and fracture process itself; and (iii) difference in environment among tests at the same load conditions and with the same material. Usually, it is not easy to separate these sources of uncertainty in a quantitative manner.

In traditional fatigue tests, the applied stress is usually fixed and treated as an independent variable, and the number of cycles to failure, N, is determined for each specimen. The scatter of fatigue life at a prescribed stress level is then quantified in a relation between the number of cycles and the proportion of failed specimens (prior to N cycles). Along this line, Sinclair and Dolan [49] performed an extensive fatigue study involving 174 identical highly polished unnotched 7075–T6 aluminum alloy specimens using six different stress levels. The results are depicted in Fig. 1.10 in the form of a histogram and show significant scatter in fatigue life. This and other investigations also indicate that greater scatter usually occurs at the lower stress levels than at the higher levels. This can be attributed to the greater

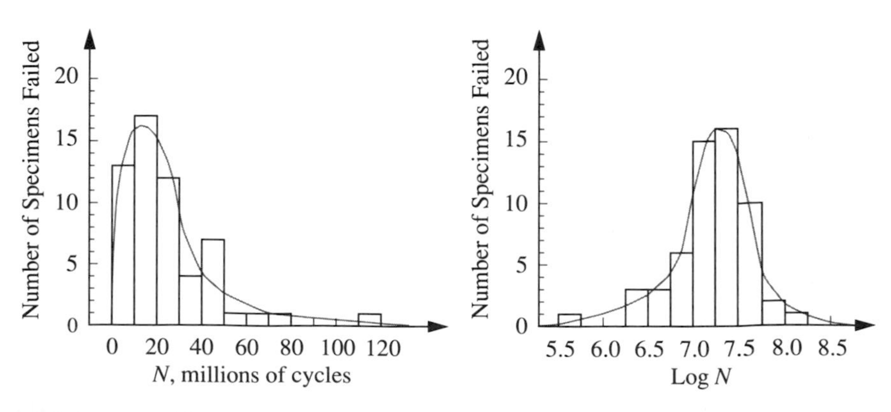

Figure 1.10. Histogram showing fatigue life distribution for 57 specimens [49].

percentage of life needed at lower stress levels to initiate small micro-cracks and then to propagate macro-cracks.

If the fatigue process is characterized by the growth of a dominant crack, then the data on crack length versus the number of cycles should be gathered and analyzed. Since fatigue crack growth is of special concern in this monograph, the data reported by Virkler et al. [54] and by Ghonem and Dore [15] will be presented next in more detail.

7.1 The Virkler Data

Virkler et al. [54] conducted a statistical investigation of fatigue crack propagation in which 68 replicate constant-amplitude crack propagation tests were conducted under identical load and environmental test conditions. Figure 1.11 shows the sample functions of crack length versus number of cycles.

The test specimens used in the experiment were 0.10 in. (2.54 mm) thick center cracked panels of 2024–T3 aluminum alloy measuring 2.2 in. (558.8 mm) long and 6 in. (152.4 mm) wide. The elements were taken from a single lot of material provided by the Aluminum Company of America. To facilitate replication of test conditions and to provide a wide range of ΔK data, constant-amplitude loading was chosen. The load signal was a si-

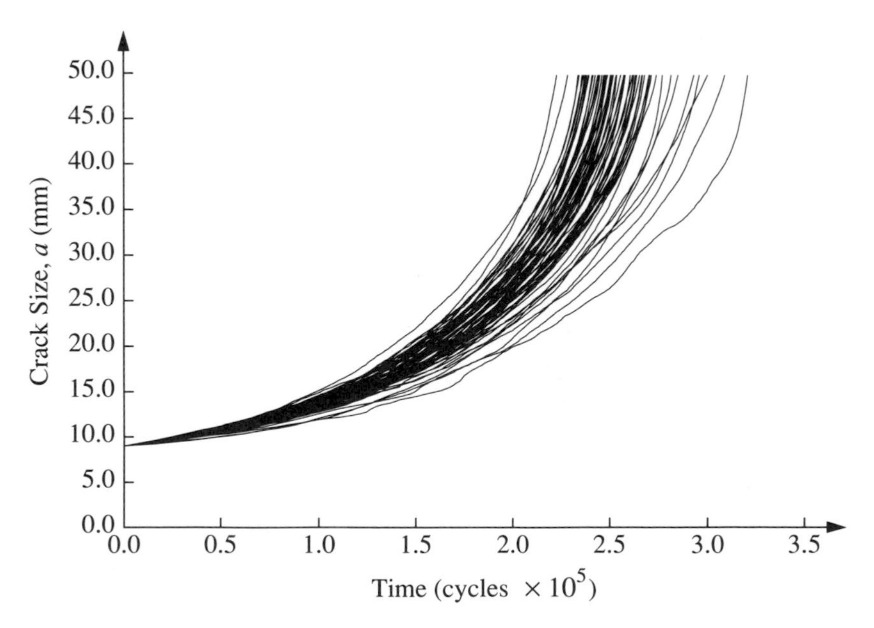

Figure 1.11. Crack length versus number of cycles for the Virkler data [54].

nusoid with a frequency of 20 Hz. Fatigue crack growth was monitored with a 150x zoom stereo microscope that was mounted on a traversing system and had a resolution of 0.001 mm. The average experimental measurement error of this setup was found to be only 0.00141 mm. The accumulated number of cycles was recorded at specified crack lengths between 9 mm and 49.8 mm. Measurements were taken every 0.2 mm in the range $9.00 \le a \le 36.20$ mm, every 0.4 mm in the range $36.20 \le a \le 44.2$ mm, and every 0.8 mm in the range $44.2 \le a \le 49.8$ mm. A total of 164 measurements were taken for each of the 68 replicate specimens.

7.2 Ghonem and Dore Data

Another, more recent collection of fatigue crack growth data has been reported by Ghonem and Dore [15]. The tests were conducted on aluminum 7075–T6 alloy rectangular specimens (320 mm × 101 mm) with thickness of 3.175 mm. The center-crack was perpendicular to the rolling direction of the sheet from which the specimens were cut. Crack length versus number of cycles data were collected at three different stress ratios. Figure 1.12 shows the plot of crack growth data (crack length a versus N number of cycles) for the case when the maximum loading amplitude was 22.79

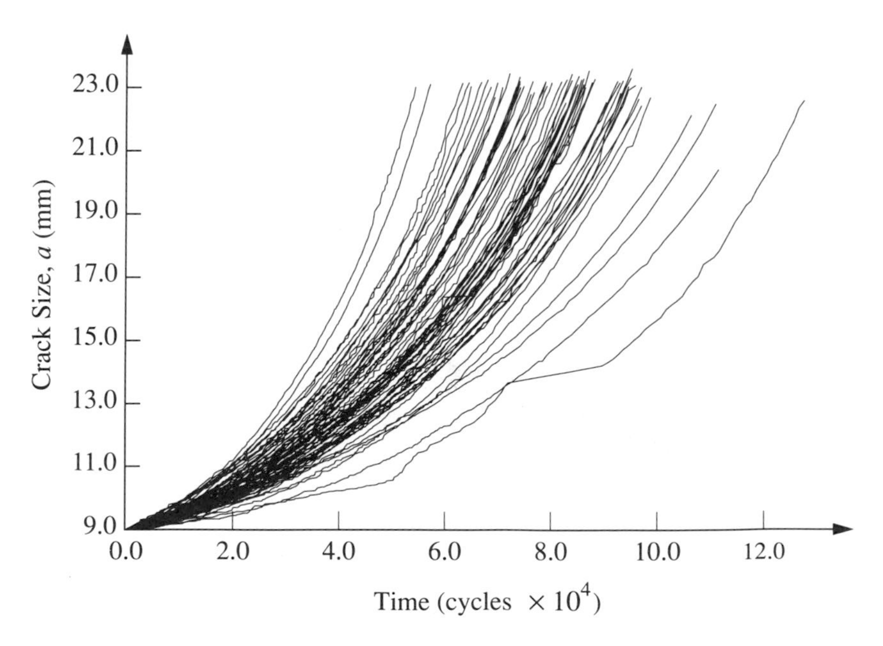

Figure 1.12. Crack length versus number of cycles for the Ghonem and Dore data [15].

KN and the stress ratio was $R = 0.6$. A total of 60 identical specimens were tested at each stress ratio.

Of various measuring techniques available (e.g., photographic technique, electrical method, ultrasonic method, etc. (see Beevers [4]), the authors adopted the photographic technique. This approach has an average experimental measurement error of 0.089 mm, which is larger than the experimental measurement error in the approach of Virkler et al. [54], but requires less manpower to take the data. In contrast to the Virkler data, which was recorded at specified crack intervals, the data reported by Ghonem and Dore was recorded at specified time intervals for crack lengths from 9 mm to 23 mm. The loading cycle was chosen to be a ramp waveform with a frequency of 10 Hz.

Remarks: As is readily seen, the data possess very significant scatter. Changes in the growth rate that occur in every test indicate, as the authors of paper [54] state, that "fatigue crack propagation is not a stable, smooth, well-ordered process." Though the material used in the tests is commonly regarded as being homogeneous, both sudden increases and decreases in the growth rate were observed as if the crack was passing through materials having different properties.

Examination of Figs. 1.11 and 1.12 leads to the following observations:

a) The crack growth sample functions are nondecreasing and they are nonreproducible, which means that no sample function can be regarded as representative of the ensemble.

b) While there is considerable intermingling of sample functions, there is no obvious condensation or clumping of sample crack growth curves.

Statistical analysis of data performed in [54] leads to the conclusion that the distribution of N is best described by the three–parameter lognormal distribution; the third parameter provides a statistical lower bound for fatigue crack growth data. It should be mentioned, however, that the analysis of the fatigue crack growth process with the use of Virkler data performed by Ditlevsen and Olesen [12] indicates that the probability distribution that is appropriate for the fatigue life (or number of cycles, N) is the inverse-Gaussian distribution. (See Section 9.2.8.)

The empirical distribution functions of the time to reach a specified crack size inferred from the data of Ghonem and Dore are provided in the review paper of Kozin and Bogdanoff [24].

Chapter II

Stochastic Approach to Fatigue: Mathematical Prerequisites

8. Introductory Remarks

The inherent scatter of fatigue data briefly discussed and visualized at the end of the last chapter, as well as the high degree of unpredictability of fatigue on the microscopic level, indicate that fatigue cannot be adequately modelled and analyzed by deterministic methods. Such a view has recently been accepted along with the conviction that the evolution of the fatigue process in engineering materials should be regarded as a random phenomenon.

Randomness of physical phenomena — its origins and its manifestation — can be interpreted and quantified in various ways. (See [62].) However, contemporary probability theory has provided the most solid and effective base for the study of regularities in various complicated and uncertain phenomena. Thus, the mathematical apparatus of probability theory — especially the methods of stochastic process theory — are believed to be appropriate for describing the behavior of fatigue processes.

As has been shown in Section 7, even under a constant-amplitude cyclic loading in well-controlled laboratory conditions, considerable statistical dispersion is found in experimental fatigue results. In addition, there are many situations in practice when structural elements are subjected to random loading (e.g., turbulent pressure, sea-wave pressure, etc.) or other random actions (e.g., environmental). Hence, when modelling fatigue in engineering structures subjected to random loading, in addition to the basic model-process $X(t)$ that characterizes the fatigue process, we require another stochastic process $S(t)$, characterizing the random external excitations or stresses.

In this chapter, we provide the basic information on probability, random variables, and stochastic processes that is particularly relevant to modelling and analysis of random fatigue. A detailed characterization of random loading processes will be presented in Chapter III.

9. Random Variables and Some Probability Distributions

9.1 Basic Concepts

To quantitatively investigate the regularities of random phenomena, one has to first introduce a mathematical model for randomness together with an appropriate measure of the possibility of occurrence of various *uncertain* outcomes of an experiment. Such a model forms a basic scheme for probability theory in which the main notions are: *sample space* (i.e., a collection of all possible outcomes of an experiment), *random event*, and *probability*.

A sample space will be denoted by Γ and its elements, the possible or elementary outcomes of an observation, by γ (i.e., $\gamma \in \Gamma$). Let $\mathcal{F}$ denote the family of subsets of Γ, termed the family of random events, on which a probability P is defined. The probability P is a function whose arguments are random events (i.e., elements of $\mathcal{F}$), such that

1. $0 \le P(A) \le 1$, for each $A \in \mathcal{F}$.
2. $P(\Gamma) = 1$.
3. For a countable collection of mutually disjoint events $A_1, A_2, \dots A_n, \dots$, in $\mathcal{F}$: $P\{\cup A_j\} = \Sigma P(A_j)$.

These three statements constitute the axioms of probability.

It is clear that in experiments on random phenomena, various outcomes, or elementary events, can occur. In most situations, they are represented by real numbers; other types of outcomes, while not originally numerical, can be made to correspond to numbers by choice of a suitable mapping. Thus, it is commonly assumed that one can assign a real number $X(\gamma)$ for each elementary event $\gamma \in \Gamma$. This leads to the concept of a random variable. The definition is the following.

A *random variable* is a real-valued function $X = X(\gamma)$, $\gamma \in \Gamma$, defined on a sample space Γ such that for every real number x there exists a probability $P\{\gamma: X(\gamma) \le x\}$. The existence of the probability of event $\{\gamma: X(\gamma) \le x\}$ ensures us (via set operations) that the probability of any finite or countably infinite combination of such events is well-defined (e.g., $P\{x_1 < X(\gamma) \le x_2\}$).

The probabilistic behavior of a random variable $X(\gamma)$ is completely and uniquely specified by its (cumulative) *distribution function* $F_X(x)$, defined as

$$F_X(x) = P\{X(\gamma) \le x\} . \tag{2.1}$$

By the definition, the distribution function always exists and is a nonnegative and nondecreasing function of the real variable x; furthermore, it is continuous from the right. From the properties of probability, it follows that

$$F_X(-\infty) = 0; \qquad F_X(+\infty) = 1, \tag{2.2}$$

and for any two real numbers a, b such that $a < b$,

$$P\{a < X \le b\} = F_X(b) - F_X(a) . \tag{2.3}$$

There are two basic classes of random variables: *discrete random variables* — taking on only a finite or countably infinite number of distinct values, and *continuous random variables* — taking on a noncountable number of values.

A discrete random variable is specified by the values it assumes, x_i, and the corresponding probabilities $p_i = P(X=x_i)$. Of course,

$$F_X(x) = \sum_{\substack{i \\ x_i \le x}} p_i , \qquad \sum_i p_i = 1. \tag{2.4}$$

Strictly speaking, a random variable $X(\gamma)$ is termed a *continuous* random variable if its probability distribution function $F_X(x)$ has a density function; i.e., for each x, there exists a nonnegative and integrable (e.g., continuous) function $f_X(x)$ such that

$$F_X(x) = \int_{-\infty}^{x} f_X(\xi)\, d\xi. \tag{2.5}$$

The function $f_X(x)$ is called the *probability density function* of the random variable $X(\gamma)$. Clearly,

$$\frac{dF_X(x)}{dx} = f_X(x) \tag{2.6}$$

and

$$\begin{aligned} & f_X(x) \geq 0, \\ & \int_a^b f_X(x)\,dx = F_X(b) - F_X(a), \\ & \int_{-\infty}^{\infty} f(x)\,dx = 1. \end{aligned} \tag{2.7}$$

The next important notion is the *mean* or *average value* of a random variable. The mean or average value of a random variable $X(\gamma)$ is defined as

$$m_X = E(X) = \langle X \rangle = \int_{-\infty}^{\infty} x dF_X(x)$$

$$= \begin{cases} \sum_i x_i p_i, & \text{for a discrete random variable} \\ \int_{-\infty}^{\infty} x f_X(x)\,dx, & \text{for a continuous random variable.} \end{cases} \tag{2.8}$$

An important class of average values is formed by the average of the powers of a random variable. These quantities are called *moments*. The kth order moment of $X(\gamma)$ is defined as

$$m_k = \langle X^k \rangle = \begin{cases} \sum_i x_i^k p_i, & \text{for a discrete random variable} \\ \int_{-\infty}^{\infty} x^k f_X(x)\,dx, & \text{for a continuous random variable.} \end{cases} \tag{2.9}$$

The *central moment* of order k of a random variable $X(\gamma)$ is the quantity

$$\mu_k = \langle [X(\gamma) - m_X]^k \rangle . \tag{2.10}$$

In particular, the second central moment is

$$\sigma_X^2 = \mu_2 = \begin{cases} \sum_i (x_i - m_X)^2 p_i, & \text{for a discrete random variable} \\ \int_{-\infty}^{\infty} (x - m_X)^2 f(x)\, dx, & \text{for a continuous random variable}, \end{cases} \tag{2.11}$$

which is known as the *variance* of $X(\gamma)$. It measures the spread or dispersion of the random variable $X(\gamma)$ about its mean. The positive square root of the variance, namely, σ_X, is called the *standard deviation* of $X(\gamma)$. The ratio,

$$v_X = \frac{\sigma_X}{|m_X|}, \qquad m_X \neq 0, \tag{2.12}$$

normalizes the spread of $X(\gamma)$ with respect to the mean, and is called the *coefficient of variation.*

Other central descriptors of a random variable are the mode and the median. The *mode* $\tilde{x}$ is the most probable value of a random variable. For continuous random variables, $\tilde{x}$ is the value for which $f_X(x)$ assumes the maximum value. The *median* x_m is the value of $X(\gamma)$ for which $F_X(x_m) = \frac{1}{2}$.

To characterize the symmetry (or lack of symmetry) of a probability distribution of random variable $X(\gamma)$, the *skewness* or *asymmetry coefficient* is defined as

$$\zeta_X^{(1)} = \frac{1}{\sigma_X^3} \langle (X - m_X)^3 \rangle = \frac{\mu_3}{\sigma_X^3}. \tag{2.13}$$

A probability distribution that is symmetric about m_X has a zero skewness coefficient (i.e., $\zeta_X^{(1)} = 0$).

The concentration of the values of $X(\gamma)$ around the mean is characterized by the *excess coefficient*

$$\zeta_X^{(2)} = \frac{1}{\sigma_X^4} \langle (X - m_X)^4 \rangle - 3 = \frac{\mu_4}{\sigma_X^4} - 3. \tag{2.14}$$

The *kurtosis coefficient* is defined as μ_4 / σ_X^4. The coefficients of skewness and excess are zero for a normal, or Gaussian, distribution (i.e., $\zeta_X^{(1)} = \zeta_X^{(2)} = 0$).

Often the joint behavior of two or more random variables is of interest. In such cases, multiple random variables and their joint probability distributions are defined and analyzed. (See Ang and Tang [55], and Soong [78].)

9.2 Probability Distributions Relevant to Fatigue

The specific probability distributions described analytically by the distribution function $F_X(x)$, or by the probability density function $f_X(x)$, can take a wide variety of forms. However, not all distributions are of equal importance. Several particular distributions have proven to be extremely useful both in theory and applications. Here, we briefly characterize the distributions that are directly relevant to fatigue life prediction.

9.2.1 Poisson Distribution

Among the distributions of discrete random variables, the Poisson distribution plays a dominant role. A discrete random variable $X(\gamma)$ taking non-negative integer values $k = 0,1,2,\ldots$ is said to have a Poisson distribution if

$$P\{X(\gamma)=k\} = \frac{\lambda^k}{k!}e^{-\lambda}, \quad k = 0, 1, 2, \ldots, \tag{2.15}$$

where λ is a positive constant characterizing the mean occurrence rate of events (per unit time). If the number of occurrences of an event in time interval of length t (e.g., in $[0, t]$) is of interest, then

$$P\{X_t=k\} = \frac{(\lambda t)^k}{k!}e^{-\lambda t}, \quad k = 0, 1, 2, \ldots . \tag{2.16}$$

The mean and variance are given by

$$\begin{aligned} \langle X_t \rangle &= m_X = \lambda t, \\ \sigma^2_{X_t} &= \lambda t. \end{aligned} \tag{2.17}$$

9.2.2 Uniform Distribution

The simplest distribution of continuous random variables is the uniform distribution. A random variable $X(\gamma)$ has a uniform distribution in the interval $[a, b]$ if

$$f_X(x) = \begin{cases} \dfrac{1}{b-a}, & a \leq x \leq b \\ 0 \; , & \text{otherwise} . \end{cases} \tag{2.18}$$

The mean and variance are given by

$$\begin{aligned} m_X &= \frac{a+b}{2} \, , \\ \sigma_X^2 &= \frac{(b-a)^2}{12} \, . \end{aligned} \tag{2.19}$$

9.2.3 Gaussian (Normal) Distribution

The most important probability distribution in theory as well as in applications is the Gaussian or normal distribution. A random variable $X(\gamma)$ with mean m_X and standard deviation σ_X has a Gaussian or normal distribution $N(m_X, \sigma)$ if its probability density function is given by

$$f_X(x) = \frac{1}{\sigma_X \sqrt{2\pi}} \exp\left[-\frac{1}{2} \left(\frac{x - m_X}{\sigma_X} \right)^2 \right], \qquad -\infty < x < \infty. \tag{2.20}$$

The cumulative distribution function $F_X(x)$ is

$$\begin{aligned} F_X(x) &= \frac{1}{\sigma_X \sqrt{2\pi}} \int_{-\infty}^{x} \exp\left[-\frac{1}{2} \left(\frac{\xi - m_X}{\sigma_X} \right)^2 \right] d\xi \\ &= \Phi\left(\frac{x - m_X}{\sigma_X} \right), \end{aligned} \tag{2.21}$$

where the function $\Phi(u)$, called the standard normal integral, is defined as

$$\Phi(u) = \frac{1}{\sqrt{2\pi}} \int_{-\infty}^{u} e^{-\xi^2/2} \, d\xi. \tag{2.22}$$

Tabulated values of $\Phi(u)$ are readily available (e.g., see [55], [78]).

It is seen that the density function (2.20) is symmetric with respect to its mean value m_X. Therefore, all odd central moments are equal to zero. In general,

$$\begin{aligned} \mu_{2k+1} &= 0, \\ \mu_{2k} &= 1 \cdot 3 \cdot \ldots \cdot (2k-1)\, \sigma_X^{2k}. \end{aligned} \tag{2.23}$$

The skewness and excess coefficient defined by Eqs. (2.13) and (2.14) are equal to zero for the Gaussian distribution.

9.2.4 Lognormal Distribution

A random variable $X(\gamma)$ has a lognormal probability distribution if $\ln X$ (the natural logarithm of X) is normal. The density function of $X(\gamma)$ is then

$$f_X(x) = \frac{1}{\zeta x \sqrt{2\pi}} \exp\left[-\frac{1}{2}\left(\frac{\ln x - \lambda}{\zeta}\right)^2\right], \qquad 0 \le x < \infty, \quad \zeta > 0, \tag{2.24}$$

where λ and ζ^2 are the mean and variance of the normal random variable $\ln X$, respectively. The mean and variance of a lognormal distribution are given in terms of the distribution parameters λ and ζ,

$$\begin{aligned} m_X &= \exp\left(\lambda + \frac{\zeta^2}{2}\right), \\ \sigma_X^2 &= m_X^2 \left(e^{\zeta^2} - 1\right). \end{aligned} \tag{2.25}$$

The cumulative distribution $F_X(x)$ is directly related to the standard normal distribution, namely,

$$F_X(x) = \Phi\left(\frac{\ln x - \lambda}{\zeta}\right). \tag{2.26}$$

From Eq. (2.25), we have

$$\begin{aligned} \lambda &= \ln(m_X) - \frac{1}{2}\zeta^2, \\ \zeta^2 &= \ln\left(1 + \frac{\sigma_X^2}{m_X^2}\right) = \ln\left(1 + \nu_X^2\right). \end{aligned} \tag{2.27}$$

9.2.5 Gamma Distribution

A random variable $X(\gamma)$ has a gamma distribution $G(a,p)$ if its probability density function takes the form

$$f_X(x) = \frac{a^p}{\Gamma(p)} x^{p-1} e^{-ax}, \quad x \geq 0, \ a > 0, \ p > 0, \tag{2.28}$$

where $\Gamma(p)$ is the well-known gamma function,

$$\Gamma(p) = \int_0^\infty x^{p-1} e^{-x} dx, \quad p > 0. \tag{2.29}$$

When p is a positive integer, $\Gamma(p) = (p-1)!$.

The cumulative distribution function is

$$F_X(x) = \int_0^x f_X(u)\, du = \frac{\Gamma(p, ax)}{\Gamma(p)}, \quad x \geq 0, \tag{2.30}$$

where $\Gamma(p, u)$ is the so-called incomplete gamma function given by

$$\Gamma(p, u) = \int_0^u \xi^{p-1} e^{-\xi} d\xi. \tag{2.31}$$

The mean and variance are

$$m_X = \frac{p}{a}, \quad \sigma_X^2 = \frac{p}{a^2}. \tag{2.32}$$

The gamma distribution serves as a useful model because of its versatility in the sense that a wide variety of shapes of the density function can be obtained by varying the values of p and a. The parameter p characterizes the shape of the distribution, whereas a is a scale parameter of the distribution. In general, the gamma density function is unimodal, with its peak at $x = 0$ for $p \leq 1$ and at $x = (p-1)/a$ for $p > 1$.

If the occurrences of an event in time are characterized by a Poisson distribution, then the time until the kth occurrence of the event is described by the gamma distribution. A special case of the gamma distribution, when p is integer, is known as the *Erlang distribution.*

9.2.6 Exponential Distribution

When $p = 1$, the gamma distribution reduces to an exponential form

$$f_X(x) = \begin{cases} ae^{-ax}, & x \geq 0, a > 0 \\ 0\ , & \text{elsewhere.} \end{cases} \tag{2.33}$$

The associated cumulative distribution function, mean, and variance are determined from Eqs. (2.30) and (2.32) by setting $p = 1$ to obtain

$$\begin{aligned} F_X(x) &= ae^{-ax}, \quad x \geq 0\ , \\ m_X = \frac{1}{a}, &\quad \sigma_X^2 = \frac{1}{a^2}\ . \end{aligned} \tag{2.34}$$

If the occurrences of events obey a Poisson distribution, then the exponential distribution describes the associated recurrence time (i.e., the time between two consecutive occurrences of the event [55]).

9.2.7 *Chi-Square (χ^2) Distribution*

Another particular case of the gamma distribution is the χ^2 distribution obtained by setting $a = \frac{1}{2}$ and $p = \frac{n}{2}$ in Eq. (2.28), where n is a positive integer. Therefore, the χ^2 distribution has the form

$$f_X(x) = \frac{1}{2^{n/2}\Gamma(\frac{n}{2})} x^{(n/2-1)} e^{-x/2}\ . \tag{2.35}$$

The parameter n is termed as the *number of degrees of freedom* of the χ^2 distribution. This distribution is best known as a distribution of a sum of the squares of n independent standard normal variables. Because of this fact, the χ^2 distribution plays an important role in statistical inference and hypothesis testing. (See χ^2–test [78].)

9.2.8 *Inverse-Gaussian Distribution*

Another distribution that might be appropriate in fatigue problems is the inverse-Gaussian distribution. Although originally related to particular problems in physics (Schrödinger and Smoluchowski, 1915) and statistics (Wald, 1944), this distribution has recently become useful in reliability theory [59] and in modelling of random fatigue. (See Chapter IV.)

A positive variable $X(\gamma)$ has an inverse-Gaussian distribution if for $x > 0$,

$$f_X(x) = f_X(x; a, b) = \sqrt{\frac{b}{2\pi}}\, x^{-3/2} \exp\left[-\frac{b(x-a)^2}{2a^2 x}\right], \tag{2.36}$$

where $a > 0, b > 0$. If we introduce a parameter $\xi = b/a$, then the density function (2.36) can be written in the following equivalent forms:

$$f_X(x;a,b) = a^{-1} f_X(\frac{x}{a}; 1, \xi) = b^{-1} f_X(\frac{x}{b}; \xi, 1) . \tag{2.37}$$

The case in which $a = 1$ is often regarded as a standard form (sometimes referred to as the standard Wald distribution). The shape of the distribution depends only on ξ; thus, ξ is termed the shape parameter.

The preceding distribution (2.37) is unimodal and skewed. However, the density function in Eq. (2.36) represents a wide class of distributions ranging from highly skewed distributions to symmetric ones as ξ varies from zero to infinity.

The cumulative distribution function $F_X(x)$ can be expressed in terms of the standard normal distribution function Φ as follows (see Shuster [75]):

$$F_X(x) = \Phi\left[\sqrt{\frac{b}{x}}\left(\frac{x}{a} - 1\right)\right] + e^{2b/a}\Phi\left[-\sqrt{\frac{b}{x}}\left(1 + \frac{x}{a}\right)\right] . \tag{2.38}$$

The first-order moments are

$$\begin{aligned} m_X &= \langle X \rangle = a , \\ m_2 &= \langle X^2 \rangle = a^2 + \frac{a^3}{b} , \\ m_3 &= \langle X^3 \rangle = a^3 + 3\frac{a^4}{6} + 3\frac{a^5}{6^2} , \\ m_4 &= \langle X^4 \rangle = a^4 + 6\frac{a^5}{b} + 15\frac{a^6}{b^2} + 15\frac{a^7}{b^3} . \end{aligned} \tag{2.39}$$

The central moments are

$$\mu_2 = \sigma_X^2 = \frac{a^3}{b}, \quad \mu_3 = 3\frac{a^5}{b^2}, \quad \mu_4 = 15\frac{a^7}{b^3} + 3\frac{a^6}{b^2} . \tag{2.40}$$

The inverse-Gaussian distribution is the first passage time distribution of the Brownian motion process (see Cox and Miller [60]); it seems that the name *inverse-Gaussian* was used for the first time just in this context by Tweedie in 1957. (See [79], [80].)

9.2.9 Extreme Value Distributions

In many applications, especially in those associated with reliability of engineering systems, there is a necessity to determine probability distribution of maximum or minimum values of a sequence of random variables, $X_1, X_2, \ldots, X_n$. (See Gumbel [65].)

Let us assume that $X_1, X_2, \ldots, X_n$ are independent and identically distributed with the same distribution $F_X(x)$. The following two random variables are of interest:

$$Y_n = \max(X_1, X_2, \ldots, X_n), \tag{2.41}$$

$$Z_n = \min(X_1, X_2, \ldots, X_n). \tag{2.42}$$

It is clear that distribution functions of the preceding variables are

$$F_{Y_n}(n) = P(Y_n \leq y) = P(\text{all } X_k \leq y) = [F_X(y)]^n, \tag{2.43}$$

$$\begin{aligned} F_{Z_n}(z) &= P(Z_n \leq z) = P(\text{at least one } X_k \leq z) \\ &= P(X_1 \leq z \cup X_2 \leq z \cup \ldots \cup X_n \leq z) \\ &= 1 - P(X_1 > z \cap X_1 > z \cap \ldots \cap X_n > z) \\ &= 1 - [1 - F_{X_1}(z)]\,[1 - F_{X_2}(z)] \ldots [1 - F_{X_n}(z)] \\ &= 1 - [1 - F_X(z)]^n. \end{aligned} \tag{2.44}$$

If the random variables X_i, $i = 1, 2, \ldots, n$ are continuous, then the corresponding density functions are given by

$$f_{Y_n}(y) = n[F_X(y)]^{n-1} f_X(y), \tag{2.45}$$

$$f_{Z_n}(z) = n[1 - F_X(z)]^{n-1} f_X(z). \tag{2.46}$$

In many situations, the number of random variables X_i is very large. Moreover, the distribution $F_X(x)$ of each X_i is often unavailable. Thus, the problem of great importance is: What are the limiting (or asymptotic) distributions of Y_n and Z_n when $n \to \infty$, and what is a possible class of distributions $F_X(x)$ for which such limiting distributions exist, are unique, and can be effectively constructed?

There are three basic types of *parent* distributions $F_X(x)$ for which the asymptotic (limiting) distributions exist.

Type I Asymptotic Distribution (of Maximum Values)

If the distribution $F_X(x)$ has the form

$$F_X(x) = 1 - e^{-g(x)}, \tag{2.47}$$

where $g(x)$ is an increasing function of x, then it is a so-called exponential-type distribution (i.e., $F_X(x)$ approaches one at least as fast as an exponential distribution). The normal, lognormal, gamma, and exponential distributions are of this type.

For exponential-type distributions, the distribution of the *maximum* of n random variates X_i, $i = 1, 2, \ldots, n$ as $n \to \infty$ is given by

$$F_Y(y) = \exp[-e^{-\alpha(y-u)}], \qquad -\infty < y < +\infty, \tag{2.48}$$

$$f_Y(y) = \alpha \exp[-\alpha(y-u) - e^{-\alpha(y-u)}]. \tag{2.49}$$

The distribution parameters are the mode u of the distribution and α, which is an inverse measure of the dispersion of the distribution. The moments are

$$m_Y = u + \frac{\delta}{\alpha}, \quad \delta \approx 0.577, \text{ Euler's constant,}$$
$$\sigma_Y^2 = \frac{\pi^2}{6\alpha^2}, \tag{2.50}$$

The preceding distribution is also termed the *double exponential*, *Gumbel*, or *Fisher–Tippet Type I* distribution. The type I distribution for minimum value can be derived in a similar manner. (See [65].)

Type II Asymptotic Distribution (of Maximum Values)

A random variate X_i is of the Pareto type if its distribution function $F_X(x)$ is given by

$$F_X(x) = 1 - ax^{-k}, \qquad a > 0, k > 0, x \geq 0. \tag{2.51}$$

For random variates of the Pareto type, the asymptotic distribution of the *maximum* of n random variates X_i, $i = 1, 2, \ldots, n$ as $n \to \infty$ has the form

$$F_Y(y) = \exp\left[-\left(\frac{u}{y}\right)^k\right]; \qquad y \geq 0, k > 0, u > 0, \tag{2.52}$$

$$f_Y(y) = \frac{k}{y}\left(\frac{u}{y}\right)^k \exp\left[-\left(\frac{u}{y}\right)^k\right], \quad y \geq 0. \tag{2.53}$$

The distribution parameters are the characteristic distribution value u (mode < u < median) and k, which is a dimensionless inverse measure of the dispersion. The moments are

$$\begin{aligned} m_Y &= u\Gamma\left(1 - \frac{1}{k}\right), \quad k > 1, \\ \sigma_Y^2 &= u^2\left[\Gamma\left(1 - \frac{2}{k}\right) - \Gamma^2\left(1 - \frac{1}{k}\right)\right], \quad k > 2. \end{aligned} \tag{2.54}$$

The moments of order l of Y do not exist for $l \geq k$; this fact complicates the estimation of the parameters u and k. It should be noticed that the parent distribution (2.51) has moments only up to the order r, where r is the largest integer less than k.

The type II asymptotic distribution of minimum values can be derived under analogous conditions. (See [65].)

Type III Asymptotic Distribution (of Minimum Values)

Because the type III asymptotic distribution for the maximum values is of limited practical interest, we present here the minimum-value distribution, which has proven to be very useful in applications.

Consider the distribution $F_X(x)$ of the random variables X_i, which is of the form

$$F_X(x) = c(x - \epsilon)^k, \quad x \geq \epsilon, c > 0, k > 0. \tag{2.55}$$

The uniform $(k = 1)$, triangular $(k = 2)$, and gamma distributions $(\epsilon = 0)$ belong to this class.

For variates distributed as in Eq. (2.55), the distribution of the *minimum* of n random variates, X_i, $i = 1, 2, \ldots, n$, as $n \to \infty$, has the form,

$$F_Z(z) = 1 - \exp\left[-\left(\frac{z-\epsilon}{w-\epsilon}\right)^k\right], \quad z \geq \epsilon, k > 0, w > \epsilon, \tag{2.56}$$

$$f_Z(z) = \frac{k}{w-\epsilon}\left(\frac{z-\epsilon}{w-\epsilon}\right)^{k-1} \exp\left[-\left(\frac{z-\epsilon}{w-\epsilon}\right)^k\right]. \tag{2.57}$$

The mean and variance are

$$m_Z = \epsilon + (w - \epsilon)\,\Gamma\left(1 + \frac{1}{k}\right),$$
$$\sigma_Z^2 = (w - \epsilon)^2\left[\Gamma\left(1 + \frac{2}{k}\right) - \Gamma^2\left(1 + \frac{1}{k}\right)\right]. \tag{2.58}$$

The distribution given by Eqs. (2.56) and (2.57) is widely known as the three-parameter *Weibull distribution*; if $\epsilon = 0$, it is termed the two-parameter Weibull distribution. For $\epsilon = 0$ and $k = 2$, the distribution (2.56)–(2.57) becomes the *Rayleigh distribution*, which has the form $(w = \sigma\sqrt{2})$,

$$f(z) = \frac{z}{\sigma^2}\exp\left(-\frac{z^2}{2\sigma^2}\right). \tag{2.59}$$

The Weibull distribution has been widely used in characterization of random fatigue lifetime. It is worth noting that the two-parameter Weibull distribution fits experimental data quite well at high stress levels. When stress levels are lower, the three-parameter distribution (with an appropriate threshold) provides a better fit of the data.

Remarks: The extreme value distributions presented earlier rely on the assumption that random variables $X_1, X_2, \ldots X_n$ are independent and identically distributed. In many practical situations, this assumption can be too restrictive. Therefore, several questions naturally come to mind: Are the formulae for the independent case still valid in the presence of dependence? Under what conditions is the extension of the preceding classical results possible? What would be the appropriate asymptotic distributions for dependent sequences when the classical results are not valid?

There are many ways of characterizing the dependence of elements in a sequence of random variables. An obvious generalization is to consider a sequence that is a Markov chain (the *past* $\{X_i,\ i<m\}$, and the *future* $\{X_i,\ i>m\}$, are independent given the *present*, X_m). This, however, would be too restrictive. An important class of dependent sequences is defined by the *m-dependence*, which means that the dependence between X_i and X_j is extended only for those X_i, X_j for which $|i-j| < m$. The correlation between X_i and X_j is also a partial measure of their dependence. Therefore, a possible dependence restriction could be given by $|\text{correlation}\,(X_i, X_j)| \le g(k)$, where $k = |i-j|$ and $g(k) \to 0$ as $k \to \infty$.

Various results from classical extreme value theory have been extended to a wide class of dependent sequences (for different dependence restrictions). A common assumption is that the sequences in question are *stationary* sequences, i.e., the distributions of $(X_{i_1}, \ldots, X_{i_n})$ and $(X_{i_1+m}, \ldots, X_{i_n+m})$ are identical for any choice of n, m, and $i_1, \ldots, i_n$.

(See Leadbetter et al. [69], Galambos [63], and Castillo [58].) For example, a stationary sequence of normal variates with the same correlation coefficients $\rho \geq 0$ between all pairs of random variables, and such that $\rho \ln(n) \to 0$ as $n \to \infty$, possesses the same asymptotic distribution (for maximum) as for the independent random variables (i.e., double exponential).

9.3 Fatigue Failure Rates

In general reliability theory, the concept of a failure rate $\mu(t)$ is often used as an alternative description of the lifetime distribution of a specimen. Let T be the random lifetime of a specimen, $F_T(t)$ the distribution function of T, and $f_T(t)$ its probability density. The nonnegative function $\mu(t)$, defined as

$$\mu(t) = \frac{f_T(t)}{1 - F_T(t)}, \tag{2.60}$$

is called the *failure intensity function* or *fatigue failure rate* (or *hazard rate*). The function $\mu(t)$ characterizes the probability that the specimen will fail between t and $t + \Delta t$ given that it has survived a time greater than t, i.e.,

$$P\{T \in (t + \Delta t) | T > t\} = \mu(t)\, \Delta t + o(\Delta t). \tag{2.61}$$

For a given $\mu(t)$, the corresponding distribution function can be easily found because Eq. (2.60) can be rewritten as

$$\frac{d}{dt} \ln[1 - F_X(t)] = -\mu(t), \tag{2.62}$$

and therefore,

$$1 - F_T(t) = \bar{F}_T(t) = \bar{F}(t_o) \exp\left\{-\int_{t_0}^{t} \mu(\tau)\, d\tau\right\}. \tag{2.63}$$

If $\bar{F}(t_0) = 1$, then

$$F_T(t) = 1 - \exp\left\{-\int_{t_0}^{t} \mu(\tau)\, d\tau\right\}, \tag{2.64}$$

$$f_T(t) = \mu(t) \exp\left\{ \int_{t_0}^{t} \mu(\tau)\, d\tau \right\}. \tag{2.65}$$

If $\mu(t) = \mu_0 = \text{constant}$, $t \geq 0$, $\mu_0 > 0$, then it is well-known that the lifetime of a specimen is exponentially distributed. However, if

$$\mu(t) = \frac{k}{w}\left(\frac{t}{w}\right)^{k-1}, \tag{2.66}$$

then the lifetime has the Weibull distribution (i.e., Eqs. (2.56)–(2.57) with $\epsilon = 0$). Many other lifetime distributions can be derived from various forms of the failure rate $\mu(t)$. Except the simplest cases, these functions have rather complicated forms. For this reason, the failure rates are often randomized (i.e., $\mu(t)$ is regarded as a realization of certain stochastic process — see Harris and Singpurwalla [66]). In the case of random failure rate, we are no longer interested in $P\{T > t \mid \mu_\gamma(t)\}$; the quantity of interest is the unconditional probability, i.e.,

$$\begin{aligned} P\{T>t\} &= \bar{F}_T(t) = \langle P\{T>t \mid \mu_\gamma(t)\}\rangle \\ &= \left\langle \exp\left\{ -\int_{t_0}^{t} \mu_\gamma(\tau)\, d\tau \right\} \right\rangle, \end{aligned} \tag{2.67}$$

where $\langle\cdot\rangle$ denotes an averaging over the ensemble of realizations of $\mu_\gamma(t)$. If $\mu_\gamma(t)$ is a stochastic process defined by a deterministic function of t with some random parameter, say, $\xi(\gamma)$, then

$$\begin{aligned} \bar{F}(t) &= \left\langle \exp\left\{ -\int_{t_0}^{t} \mu(\tau; \xi(\gamma))\, d\tau \right\} \right\rangle \\ &= \int \left\{ \exp\left[-\int_{t_0}^{t} \mu(\tau; \xi(\gamma))\, d\tau \right] \right\} f(\xi)\, d\xi, \end{aligned} \tag{2.68}$$

where $f(\xi)$ is probability density of random variable $\xi(\gamma)$, and the integration is extended over the range of values of $\xi(\gamma)$. If $\mu_\gamma(t)$ is regarded as a realization of a stochastic process with independent increments, then a wide class of lifetime distributions can be derived by use of Eq. (2.67). (See Antelman and Savage [56].) An extension of this idea to analysis of random fatigue process is described in [76]

9.4 Entropy and Informational Content

A result of an experiment associated with a random phenomenon cannot be predicted with certainty before its observation; a characteristic feature of any random phenomenon is its uncertainty. This uncertainty (or indeterminacy) can often be estimated by qualitatively comparing uncertainties resulting from different situations. However, in the analysis of random phenomena (especially those found in engineering systems), it is desirable to evaluate the uncertainty quantitatively. A convenient measure of uncertainty is given as a function of the experimental outcome probability and is termed *entropy.* The entropy of a random variable that models an experimental outcome is a measure of the indeterminacy (or randomness) of the experiment's outcome before the experiment has been performed. As we shall see shortly, the entropy of a discrete random variable can be interpreted as an average number of *bits* necessary to differentiate its possible values. Closely related to entropy is the concept of *information.* It turns out that the amount of information contained in a random variable, or the amount of mutual information between two random variables, can be defined in terms of entropy.

Let $X(\gamma)$ be a discrete random variable assuming values x_i with probabilities $P\{X(\gamma)=x_i\} = p_i$, $i = 1, 2, \ldots, n$. The quantity $H(X)$, defined as

$$H(X) = -\sum_{i=1}^{n} p_i \log p_i, \tag{2.69}$$

is called the *entropy of a discrete random variable* $X(\gamma)$.

The negative sign in Eq. (2.69) makes the entropy nonnegative and allows the logarithm to be taken with arbitrary base greater than one. When a base 2 logarithm is used, the unit of entropy is called a *bit* (binary digit). When the natural logarithm (i.e., base e) is used, the unit is called a *nit.*

Let us consider a random variable $X(\gamma)$ that assumes two values, x_1, x_2, with equal probabilities of $\frac{1}{2}$. In the case of the base 2 logarithms, we have

$$H(X) = -\left[\frac{1}{2}\log_2\left(\frac{1}{2}\right) + \frac{1}{2}\log_2\left(\frac{1}{2}\right)\right] = 1 \text{ bit}.$$

Let us notice that the entropy of a discrete random variable depends only on the number of values and their probabilities and does not depend on the values themselves.

Let (X, Y) be a two-dimensional random vector assuming values (x_i, y_i) with probabilities p_{ik}, $i = 1, 2, \ldots, n$; $k = 1, 2, \ldots, m$. The entropy of random vector (X, Y) is defined as

$$H(X, Y) = -\sum_{i=1}^{n} \sum_{k=1}^{m} p_{ik} \log p_{ik}. \tag{2.70}$$

The entropy of a continuous random variable $X(\gamma)$ with probability density $f(x)$ is given by

$$\begin{aligned} H(X) &= -\int_{-\infty}^{\infty} f(x) \log f(x)\, dx \\ &= \langle [-\log f(X)] \rangle. \end{aligned} \tag{2.71}$$

It can be easily calculated that entropy of the uniform distribution on interval $\lfloor a, b \rfloor$ is

$$H(X) - \log(b - a), \tag{2.72}$$

whereas the entropy of Gaussian random variable in base e is equal to

$$H(X) = \frac{1}{2} \ln (2\pi\, \mathrm{e} \sigma_X^2). \tag{2.73}$$

Note that the entropy of the uniform and Gaussian distributions are monotonic functions of the interval $a - b$ and variance σ^2, respectively. This implies that the entropy of a continuous random variable can assume both positive and negative values.

The entropy of an n-dimensional random vector $\mathbf{X} = [X_1, \ldots, X_n]$ with the probability density $f(x_1, \ldots, x_n)$ is defined as

$$\begin{aligned} H(X) &= -\int_{-\infty}^{\infty} \cdots \int_{-\infty}^{\infty} f(x_1, \ldots, x_n) \log f(x_1, \ldots, x_n)\, dx, \ldots, dx_n \\ &= \langle [-\log f(\mathbf{X})] \rangle. \end{aligned} \tag{2.74}$$

The entropy of an n-dimensional Gaussian vector $\mathbf{X}$ in base e is given by

$$H(X) = \ln \sqrt{(2\pi\, \mathrm{e})^n |\mathbf{K}|}. \tag{2.75}$$

The conditional entropy of a random variable $X(\gamma)$ with respect to random variable $Y(\gamma)$ is the quantity

$$H(X|Y) = -\int_{-\infty}^{\infty} f(x|y) \log f(x|y)\, dx. \tag{2.76}$$

The preceding quantity depends on the values of the random variable $Y(\gamma)$, so it is itself a random variable.

The average conditional entropy of $X(\gamma)$ with respect to $Y(\gamma)$ is defined as

$$\begin{aligned} \overline{H}_Y(X) &= \langle H(X|Y) \rangle = \int_{-\infty}^{\infty} H(X|Y) f(y)\, dy \\ &= -\int_{-\infty}^{\infty} f(x,y) \log f(x|y)\, dx dy\ . \end{aligned} \tag{2.77}$$

An analogous definition of conditional entropy holds for discrete random variables. For both continuous and discrete random variables,

$$\begin{aligned} H(X,Y) &= H(X) + \overline{H}_X(Y) = H(Y) + \overline{H}_Y(X) \\ &\leq H(X) + H(Y)\,, \end{aligned} \tag{2.78}$$

where equality holds if and only if the random variables are independent.

The mutual information $I(X,Y)$ between random variables $X(\gamma)$ and $Y(\gamma)$ is defined as

$$I(X,Y) = H(X) - \overline{H}_Y(X)\,. \tag{2.79}$$

From Eq. (2.78), one then obtains

$$I(X,Y) = H(X) + H(Y) - H(X,Y)\,. \tag{2.80}$$

For continuous random variables,

$$I(X,Y) = \int_{-\infty}^{\infty}\int_{-\infty}^{\infty} f(x,y) \log \frac{f(x,y)}{f(x)f(y)}\, dx dy. \tag{2.81}$$

The basic properties of the mutual information are:

a) $I(X,Y) \geq 0$, equality holds if and only if $X(\gamma)$ and $Y(\gamma)$ are independent.
b) $I(X,Y) = I(Y,X)$.
c) If $Z = g(X)$ is a one-to-one mapping, then $I(X,Y) = I(Z,Y)$.

The concepts of entropy and informational content (indirectly provided by observation of other random variables) are useful in statistical inference and, specifically, in comparing various models and probability distributions. (See Chapter VII.)

10. Stochastic Processes

10.1 Basic Concepts

In many applications, one has to deal with random quantities that depend upon a certain deterministic parameter. One of the well-known examples of such phenomena is *Brownian motion*. Each coordinate of the Brownian particle is a random variable that depends on time. Other examples are the ground motion of a fixed spatial point during an earthquake (which at each instant of time is a random variable), wind pressure acting on a structure, sea wave excitation, etc.

To create a mathematical model of phenomena similar to those mentioned before, the concept of a random function has been introduced. Most often, the parameter (or dependent variable) in the random functions is time, and the concepts of stochastic processes are commonly used.

Let T be a set of real numbers and let $t \in T$, where t has the meaning of time. A *stochastic process* $X(t)$ is a family of random variables $\{X_t(\gamma),\ t \in T,\ \gamma \in \Gamma\}$ depending upon the parameter t and defined on the probability space $(\Gamma, \mathcal{F}, P)$. One can also say that a stochastic process $X(t)$ is a function that maps the index set T into the space S of random variables defined on $(\Gamma, \mathcal{F}, P)$.

Since a stochastic process is a family of random variables, its specification is similar to that for random vectors; the differences are associated with the fact that the number of random variables may now be (countably or uncountably) infinite. It turns out, however, that an infinite number of random variables can be described by means of finite-dimensional distributions. For an arbitrary finite set of t-values, say, $\{t_1, t_2, \ldots, t_n\}$, the random variables $X(t_1), X(t_2), \ldots, X(t_n)$ have a joint n-dimensional distribution with the distribution function,

$$\begin{aligned} &F_{t_1, t_2, \ldots, t_n}(x_1, x_2, \ldots, x_n) \\ &\qquad = P\{X(t_1) \le x_1,\ X(t_2) \le x_2,\ \ldots,\ X(t_n) \le x_n\}. \end{aligned} \tag{2.82}$$

The function $F_{t_1, t_2, \ldots, t_n}(x_1, x_2, \ldots, x_n)$ is called the *n-dimensional distribution function* of the random process $X(t)$. The collection of all joint distribution functions for $n = 1, 2, \ldots$ at all possible time instants t_i constitutes the family of finite-dimensional distributions of the process

$X(t)$; this family of distributions (for all finite sets $\{t_1, t_2, \ldots, t_n\}$ of $t \in T$) characterizes the process $X(t)$. The associated joint density function $f_{t_1, t_2, \ldots, t_n}(x_1, x_2, \ldots, x_n)$ is called the *n-dimensional probability density function* of the process $X(t)$.

A stochastic process can also be defined in another way. For every fixed elementary event γ in the given probability space Γ, $X(t)$ — or, more explicitly, $X(t, \gamma)$ — becomes a function of t, defined for all $t \in T$. This function $x(t) = X(t, \gamma)$, for a fixed $\gamma \in \Gamma$, describes a particular realization of the stochastic process and is called a *realization, trajectory,* or *sample function* of the process. The sample function $x(t)$ may be regarded as a *point* in the space $\mathfrak{X}$ of all finite real-valued functions of $t \in T$; $\mathfrak{X}$ is called the *sample function space* of the process $X(t)$. (See Fig. 2.1.) Thus, the stochastic process $X(t)$, or $X(t, \gamma)$, can be defined as a *generalized random variable* (or *random element*), i.e., as a function mapping every point $\gamma \in \Gamma$ into a point of the space $\mathfrak{X}$. According to such an interpretation, a stochastic process is specified by characterizing the probability (or probability measure) defined on the sample function space $\mathfrak{X}$. A possible way to construct such a probability measure is associated with the concept of the characteristic functionals of the process.

Like random variables, stochastic processes can be conveniently described by their moments; the simplest characteristic is the *mean* or *average value,* $m_X(t)$. It is defined as a function that, for each t, is equal to the mean value of the corresponding random variable, i.e.,

$$m_X(t) = \langle X(t) \rangle, \tag{2.83}$$

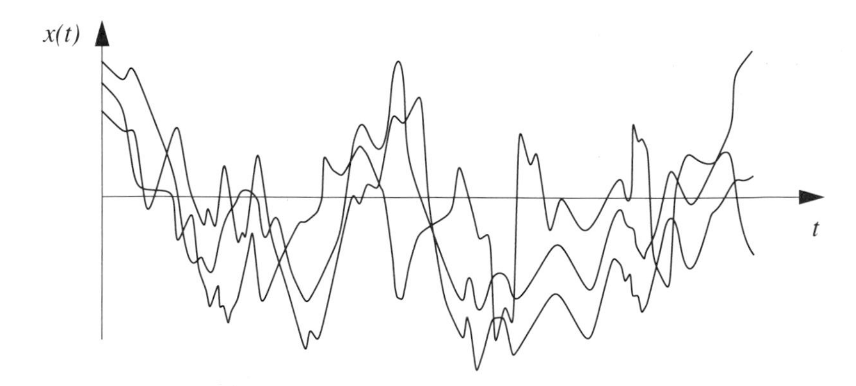

Figure 2.1. Realizations of a stochastic process.

where $\langle \cdot \rangle$ denotes the ensemble average. The average value of a real stochastic process $X(t)$ is expressed using the one-dimensional density function of the process $f_t(x)$ as follows:

$$m_X(t) = \int_{-\infty}^{\infty} x \, dF_t(x) = \int_{-\infty}^{\infty} x \, f_t(x) \, dx. \tag{2.84}$$

The variance of the process $X(t)$ is defined by

$$\sigma_X^2(t) = \langle [X(t) - m_X(t)]^2 \rangle. \tag{2.85}$$

One of the basic quantities of interest both in theory and in applications is the covariance function of the stochastic process. The *covariance function* of a real stochastic process $X(t)$ is defined as

$$K_X(t_1, t_2) = \langle [X(t_1) - m_X(t_1)] [X(t_2) - m_X(t_2)] \rangle. \tag{2.86}$$

It is easily seen that $\sigma_X^2(t) = K_X(t, t)$. The function $\langle X(t_1) X(t_2) \rangle$ is often called the *correlation function*. A more complete characterization of the stochastic process can be provided by the higher-order moments, for example by the function

$$K_X(t_1, \ldots, t_2) = \langle [X(t_1) - m_X(t_1)] \ldots [X(t_n) - m_X(t_n)] \rangle, \tag{2.87}$$

which is sometimes called the *moment function of the nth order*.

Often, the simultaneous behavior of n stochastic processes $X_1(t), \ldots, X_n(t)$ is of interest. One can then introduce an n-dimensional stochastic vector process $\mathbf{X}(t) = [X_1(t), \ldots, X_n(t)]$. The mean value of such a process is defined as: $\mathbf{m_X}(t) = [m_{X_1}(t), m_{X_2}(t), \ldots, m_{X_n}(t)]$. The component processes $X_i(t)$ may be statistically related to each other; their mutual dependence is conveniently characterized by the covariance matrix $\mathbf{R}(t_1, t_2) = \{R_{X_iX_j}(t_1, t_2)\}$, $i, j = 1, 2, \ldots, n$, where

$$R_{X_iX_j}(t_1, t_2) = \langle [X_i(t_1) - m_{X_i}(t_1)] [X_j(t_2) - m_{X_j}(t_2)] \rangle \tag{2.88}$$

is said to be the *cross-covariance function* of the processes $X_i(t)$ and $X_j(t)$. Obviously, $R_{X_iX_i}(t_1, t_2) = K_{X_i}(t_1, t_2)$.

A stochastic process $X(t)$, $t \in T$, is called a *Gaussian* or *normal stochastic process* if all of its finite-dimensional distributions are Gaussian; i.e., its joint probability density function has the form

$$f_{t_1, \ldots, t_n}(x_1, \ldots, x_n)$$
$$= \frac{1}{(2\pi)^{n/2}\sqrt{|\mathbf{K}|}} \exp\left(-\frac{1}{2|\mathbf{K}|}\sum_{i,j=1}^{n} |K_{ij}|[x_i - m_X(t_i)][x_j - m_X(t_j)]\right), \tag{2.89}$$

where $x_i = X(t_i)$, $|\mathbf{K}| \neq 0$, is the determinant of the covariance matrix $\mathbf{K}_X(t_i, t_j)$, $i, j, = 1, 2, \ldots, n$, of the process $X(t)$, and $|K_{ij}|$ is the cofactor of the element K_{ij} of the matrix $\mathbf{K}_X(t_i, t_j)$. Hence, the n-dimensional distributions of a Gaussian process are completely specified by the mean $m_X(t)$ and the covariance function $K_X(t_1, t_2)$. The moment functions of higher order can be expressed in terms of $K_X(t_1, t_2)$. For example,

$$\begin{aligned} K_X(t_1, t_2, t_3, t_4) &= K_X(t_1, t_2)K_X(t_3, t_4) + K_X(t_1, t_3)K_X(t_2, t_4) \\ &\quad + K_X(t_1, t_4)K_X(t_2, t_3). \end{aligned} \tag{2.90}$$

10.2 Stationary Processes

In many fields of application, we encounter random processes that are invariant under a translation of time. A useful model of such phenomena is commonly called a *stationary stochastic process*.

A stochastic process $X(t)$, $t \in T$, is said to be *strictly stationary* if the family of all finite-dimensional distributions (2.82) remains invariant under an arbitrary translation of the time parameter; i.e., for each n and for an arbitrary τ such that $t_i + \tau \in T$, we have

$$F_{t_1, t_2, \ldots, t_n}(x_1, x_2, \ldots, x_n) = F_{t_1+\tau, t_2+\tau, \ldots, t_n+\tau}(x_1, x_2, \ldots, x_n). \tag{2.91}$$

Given a physical process, it is often quite difficult to determine whether it is strictly stationary, since the equality in Eq. (2.91) has to hold for all n. To make things easier, a wider class of stationary processes is introduced.

A stochastic process $X(t)$, $t \in T$, is called a *weakly stationary* process (or, a *wide-sense stationary* process) if

$$\begin{aligned} \langle X(t) \rangle &= m_X(t) = m_X = \text{constant} < \infty, \\ K_X(t_1, t_2) &= K_X(t_2 - t_1) = K_X(\tau), \quad \tau = t_2 - t_1. \end{aligned} \tag{2.92}$$

Clearly, a strictly stationary process whose second moment is finite is also weakly stationary. The converse statement is not true in general. An exception is the Gaussian process, since it is completely characterized by its moments of first and second order. Therefore, a weakly stationary Gaussian process is also strictly stationary.

A stochastic vector process $\mathbf{X}(t) = [X_1(t), X_2(t), \ldots, X_n(t)]$ is defined as *weakly stationary* if its mean is constant and all elements of its covariance matrix $\mathbf{R}(t_1, t_2)$ depend only on the difference $\tau = t_2 - t_1$. From this definition, it follows that the stationarity of a vector stochastic process is not assured by the stationarity of its components; the component processes also have to be stationarily correlated.

Stationary processes are of great practical importance. This is mainly due to their regularity. The estimation of the statistical characteristics of such processes from experimental data is much simpler than in the case of nonstationary processes. Furthermore, for stationary processes, there exists the apparatus of spectral analysis, analogous to harmonic analysis of deterministic functions. The spectral method follows from the *Bochner–Khincin theorem* [61], which states that a function $K_X(\tau)$ can be a correlation function of a stationary (and mean-square continuous) stochastic process if and only if it has the representation

$$K_X(\tau) = \int_{-\infty}^{+\infty} e^{i\omega\tau} dG_X(\omega), \tag{2.93}$$

where $G_X(\omega)$ is termed the spectral distribution function and is real, nondecreasing, bounded, and continuous from the left.

If $G_X(\omega)$ is absolutely continuous, i.e., there exists a function $g_X(\omega)$ such that

$$G_X(\omega) = \int_{-\infty}^{\omega} g_X(\xi)\, d\xi, \tag{2.94}$$

$$g_X(\omega) = \frac{dG_X(\omega)}{d\omega}, \tag{2.95}$$

then the function $g_X(\omega)$ is called the *spectral density* of the process $X(t)$. Equation (2.93) can be rewritten in terms of the spectral density function as

$$K_X(\tau) = \int_{-\infty}^{\infty} e^{i\omega\tau} g_X(\omega)\, d\omega, \tag{2.96}$$

and (provided $\int_{-\infty}^{\infty} |K_X(\tau)|\, d\tau < \infty$) we have

$$g_X(\omega) = \frac{1}{2\pi}\int_{-\infty}^{\infty} e^{-i\omega\tau} K_X(\tau)\, d\tau. \tag{2.97}$$

It is easily seen that

a) $g_X(\omega) \geq 0$.
b) $\sigma_X^2 = K_X(0) = \int_{-\infty}^{\infty} g_X(\omega)\, d\omega$.
c) For real processes: $g_X(\omega) = g_X(-\omega)$.

Figure 2.2 shows a typical correlation function and the corresponding spectral density of a stochastic process $X(t)$.

A useful method of specifying stochastic processes is to characterize them in terms of a deterministic function of time with random variables as parameters; that is,

$$X(t) = g(t; X_1(\gamma), \ldots, X_n(\gamma)), \tag{2.98}$$

where the functional form of g is given and the random variables $X_1(\gamma), \ldots, X_n(\gamma)$ have specified probabilistic properties. A common example is a random harmonic oscillation,

$$X(t) = A\cos\omega t + B\sin\omega t, \tag{2.99}$$

where A and B are random variables and ω is constant. If A and B are uncorrelated random variables with zero mean and equal variance, then the process in Eq. (2.99) is weakly stationary. If additionally, A and B are Gaussian, then this process is also Gaussian. The stochastic process in Eq. (2.99) can be equivalently represented as

$$X(t) = R\cos(\omega t + \Phi), \tag{2.100}$$

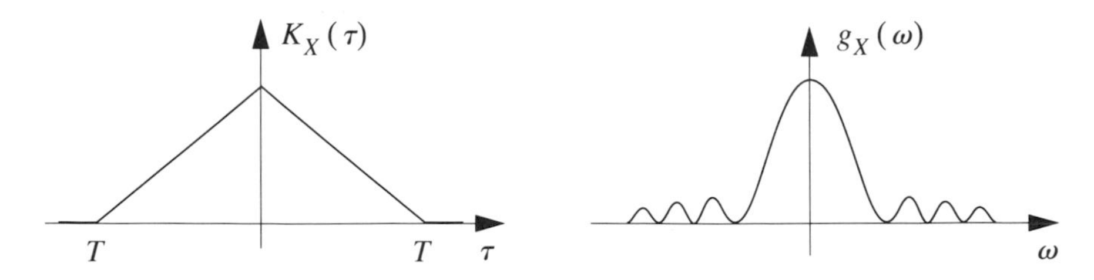

Figure 2.2. Correlation function and corresponding spectral density of a stationary process.

where $R = \sqrt{A^2+B^2}$, and $\Phi = \arctan(A/B)$. In the Gaussian case, R and Φ are independent, R having the Rayleigh distribution and Φ being uniformly distributed on the interval $(0, 2\pi]$.

It is worth noting that the process $Y(t)$, termed a *periodic signal in noise* and defined by

$$Y(t) = r\cos(\omega t+\varphi) + X(t), \tag{2.101}$$

where r, ω, and φ are deterministic constants and $X(t)$ is a stationary Gaussian stochastic process, is Gaussian but nonstationary.

In the analysis of various practical problems employing stationary stochastic processes, it is important to know the range of frequencies that contain the most significant power in a random signal. This leads to the concepts of broad-band and narrow-band stationary processes.

Usually, one says that a stationary process $X(t)$ is *broad-* or *wide-band* if its spectral density function has significant values over a wide range of frequencies. If the process $X(t)$ plays the role of an excitation to a dynamical system, and its band of *significant* frequencies is large in comparison with the frequency band of the system in question, then the process $X(t)$ is said to be a broad-band excitation. A process with the reverse property (i.e., $g_X(\omega)$ has significant values over a narrow band of frequencies) is called a *narrow-band* process. Figure 2.3 depicts trajectories and spectral densities of narrow- and broad-band processes. We see that the narrow-band

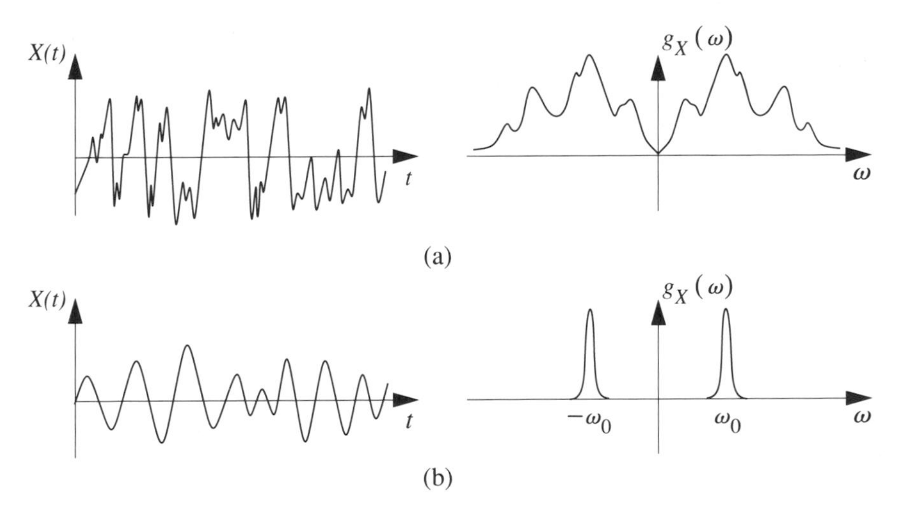

Figure 2.3 Trajectories and spectral densities of (a) broad-band process; (b) narrow-band process.

process is characterized by its approximate constant period or frequency; the frequency content of the wide-band process is more complex.

A common idealization of a broad-band process is to assume that $g_X(\omega) = g_o$ = constant for all frequencies. Such a process is called a *white noise* process, which will be denoted by $\xi(t)$. This implies that

$$g_\xi(\omega) = \frac{1}{2\pi}\int_{-\infty}^{\infty} K_\xi(\tau)\, e^{-i\omega\tau} d\tau = g_o = \frac{1}{2\pi}\nu, \tag{2.102}$$

where the constant ν is called the intensity of the white noise. Such a process does not exist in the classical sense because Eq. (2.102) is only compatible with the following correlation function:

$$K_\xi(\tau) = \nu\delta(\tau), \tag{2.103}$$

where $\delta(\tau)$ is the Dirac delta function. Equation (2.102) implies that a white noise process has an infinite variance (i.e., $K_\xi(0) = \sigma_\xi^2 = \infty$). Thus, white noise is physically unrealizable. Despite these defects, white noise provides a useful approximation of real processes with very short correlation time and is widely used in applications. The use of a white noise process in stochastic analysis resembles in many ways the use of the Dirac delta function in deterministic theory of dynamical systems. A rigorous definition of white noise can be given within the theory of generalized stochastic processes.

Spectral characteristics are commonly used in the analysis of stochastic processes. The spectral content of the process is often characterized by several parameters. The most important of them — the *spectral moments* λ_k and the *regularity factor* α — are defined as follows:

$$\lambda_k = \int_{-\infty}^{+\infty} \omega^k g_X(\omega)\, d\omega , \tag{2.104}$$

$$\alpha = \frac{\lambda_2}{\sqrt{\lambda_0 \lambda_4}}. \tag{2.105}$$

We see that spectral moments, defined by an integral over infinite range, may or may not be finite. The moment λ_{2k} is finite if and only if $K_X(\tau)$ possesses a derivative of order $2k$ at $\tau = 0$. It can be easily shown that [61]

$$\lambda_{2k} = (-1)^k \left. \frac{d^{2k} K_X(\tau)}{d\tau^{2k}} \right|_{\tau = 0}, \tag{2.106}$$

where it is clear that $\lambda_0 = K_X(0) = \sigma_X^2$.

The parameter α characterizes the bandwidth of the process. If the process is narrow-band, $\alpha \approx 1$. The parameters λ_k and α are especially useful in characterizing the extreme values of a stochastic process. In Chapter III, the rate of positive zero-crossings as well as the rate of maxima above the zero level are shown to be functions of spectral moments.

10.3 Stochastic Derivatives and Integrals

In applications of stochastic process theory to the analysis of dynamic phenomena, one often has to deal with differential equations for random functions. To be able to interpret such equations correctly, it is necessary to define the concepts of a derivative and integral of a stochastic process. These two concepts are of primary interest in applications.

If almost all sample functions of a process $X(t)$ (i.e., all except events with zero probability) are differentiable (integrable), then one usually says that the process is *differentiable (integrable) with probability one*. Knowledge of the analytic properties of sample functions of the stochastic process under consideration is of considerable interest in many applied problems. It turns out, however, that in correlational analysis of stochastic processes (dealing only with the first and second moments of a process), the concept of a derivative and integral in the mean-square sense is very natural. These concepts are defined in the following way. A stochastic process $X(t)$ with finite second-order moments is said to have a *mean-square derivative* $\dot{X}$ at $t \in T$ if

$$\lim_{h \to 0} \left\langle \left| \frac{X(t+h) - X(t)}{h} - \dot{X}(t) \right|^2 \right\rangle = 0, \quad t + h \in T. \tag{2.107}$$

Analogously, the *mean-square integral* of a stochastic process $X(t)$ is defined by the relation

$$\lim_{n \to \infty} \left\langle \left| \sum_{i=1}^{n} X(\tau_i)(t_{i+1} - t_i) - \int_a^b X(t)\, dt \right|^2 \right\rangle = 0 \tag{2.108}$$

for each partition of the interval $[a, b]$ and for arbitrary $\tau_i \in [t_i, t_{i+1}]$.

It is possible to establish the properties of mean-square derivatives and integrals and the rules of their operation that prove to be analogous to those in calculus of ordinary functions. (See [61].) Here, we shall only give those properties that occur most commonly in applications.

If the mixed derivative $\partial^2 K_X / \partial t_1 \partial t_2$ of the correlation function $K_X(t_1, t_2)$ exists, then $X(t)$ has a mean-square derivative $X(t)$, and

$$K_{\dot{X}}(t_1, t_2) = \frac{\partial^2}{\partial t_1 \partial t_2} K_X(t_1, t_2) \ . \tag{2.109}$$

If a stochastic process $X(t)$ is stationary, then its derivatives are also stationary, and

$$K_{\dot{X}}(\tau) = -\frac{d^2 K_X(\tau)}{d\tau^2}. \tag{2.110}$$

The spectral density of the derivative process $\dot{X}(t)$ is given by

$$g_{\dot{X}}(\omega) = \omega^2 g_X(\omega), \tag{2.111}$$

and in general,

$$g_{X^{(n)}}(\omega) = \omega^{2n} g_X(\omega), \tag{2.112}$$

where $X^{(n)}$ denotes the nth derivative of X with respect to time.

A stochastic process $X(t)$ with correlation function $K_X(t_1, t_2)$ is mean-square integrable over the interval $[a, b]$ if and only if the ordinary double Riemann integral

$$\int_a^b \int_a^b K_X(t_1, t_2)\, dt_1 dt_2$$

exists (and is finite). Let $\varphi(t)$ be an ordinary function defined and integrable on the interval $[a, b]$, and let

$$Y(t) = \int_{t_o}^{t} \varphi(s) X(s)\, ds. \tag{2.113}$$

Then

$$K_Y(t_1, t_2) = \int_{t_o}^{t_1}\int_{t_o}^{t_2} \varphi(s_1)\,\varphi(s_2)\,K_X(s_1, s_2)\,ds_1 ds_2. \tag{2.114}$$

Remarks: It is well-known that differential equations provide very convenient models for numerous physical processes. Additionally, as was seen in Chapter I, the fatigue process can be modelled by means of differential equations. However, if we wish to take into account the complexity and uncertainty associated with the processes considered, then we come to models expressed by *stochastic differential equations*; that is, by differential equations for stochastic processes.

The differential equation for a stochastic process $Y(t)$,

$$\frac{dY(t)}{dt} = F[Y(t), X(t, \gamma), t], \quad Y(t_0) = Y_0, \tag{2.115}$$

where F is deterministic function of its arguments, and $X(t, \gamma)$ is a given stochastic process, represents a stochastic differential equation. The initial condition for Eq. (2.115), Y_o, can be random or deterministic. If the random processes occurring in Eq. (2.115) are sufficiently regular, then the main problems are similar to those in the classical theory of differential equations; in this case, Eq. (2.115) can be interpreted as a family of equations for the individual sample functions. Of course, introducing random elements into differential equations also generates new problems; of most interest are the questions concerned with the probabilistic properties of the solution process.

If Eq. (2.115) involves highly irregular stochastic processes like a white noise or Brownian motion process, then we have a special class of stochastic differential equations that requires a specific theory. The most common representation of such equations is the *Itô stochastic differential equation,* whose usual form is

$$dY(t) = m[Y(t), t]\,dt + \sigma[Y(t), t]\,dW(t), \tag{2.116}$$

where $Y(t)$ is the unknown stochastic process, $W(t)$ is the Wiener process (defined in the next section), and $m(y, t)$ and $\sigma(y, t)$ are given functions defined for $t \in [t_o, T]$, $y \in (-\infty, \infty)$.

Since the realizations of the Wiener process are nondifferentiable (and are, with probability 1, of unbounded variation in any interval), the integration of Eq. (2.116) requires a special definition of the Stieltjes-type integral with respect to the Wiener process $W(t)$. Such an integral is known as the *Itô stochastic integral* and plays a fundamental role in the analysis of equations of the form (2.116). The Itô stochastic differential equations serve as models of dynamical systems with rapidly varying random excitations; for example, with excitations being white noises. The important feature of Itô stochastic equations is that, under quite general conditions, the solutions are diffusion Markov processes. For further information, the reader is referred to [77]. Stochastic differential equation models of fatigue will be discussed in Chapter VI.

10.4 Markov Diffusion Processes

An important class of stochastic processes possessing a well-developed theory, as well as a very broad range of applications, is the class of Markov processes. One can say that in a Markov process, the *future* is independent of the *past* when we know the *present*. More exactly, a stochastic process $X(t)$, $t \in T$, is said to be a *Markov process* if for an arbitrary set of times $t_1 < t_2 < \ldots < t_n$, the conditional distribution of $X(t_n)$ when $X(t_1), \ldots, X(t_{n-1})$ are given depends only on $X(t_{n-1})$, i.e.,

$$\begin{aligned} P\{X(t_n) < x_n | X(t_{n-1}) < x_{n-1}; \ \ldots \ ; X(t_2) < x_2; X(t_1) < x_1\} \\ = P\{X(t_n) < x_n | X(t_{n-1}) < x_{n-1}\}. \end{aligned} \tag{2.117}$$

If the distribution of the initial state is known, then the Markov process is completely specified by the following conditional distribution function,

$$F(y, t|x, s) = P\{X(t) < y | X(s) = x\}, \tag{2.118}$$

which is commonly known as the *transition probability function* of the Markov process $X(t)$.

Let us restrict our attention to *continuously valued Markov processes*. The basic class of such processes is formed by diffusion Markov processes. Without going into details, one can say that *diffusion processes* are the processes for which the probability of any substantial change within a small time interval Δt is very small (of order less than Δt) and there exist functions $a(x, t)$ and $b(x, t)$ characterizing, respectively, the average tendency in the evolution of the process $X(t)$ within an infinitesimal time interval $(t, t + \Delta t)$ and the mean square of the change in the process within the time interval $(t, t + \Delta t)$. For a diffusion Markov process, its transition probability function $F(y, t|x, s)$ and the corresponding density $p(y, t|x, s)$ satisfy certain partial differential equations of the parabolic type. One of these equations is used particularly often in applications; it is the *Fokker–Planck–Kolmogorov* equation,

$$\frac{\partial p}{\partial t} + \frac{\partial}{\partial y}[a(y, t) p(y, t|x, s)] - \frac{1}{2}\frac{\partial^2}{\partial y^2}[b(y, t) p(y, t|x, s)] = 0. \tag{2.119}$$

For a vector Markov diffusion process $\mathbf{X}(t) = [X_1(t), \ldots, X_n(t)]$, the Fokker–Planck–Kolmogorov equation has the form,

$$\frac{\partial p(\mathbf{y}, t|\mathbf{x}, s)}{\partial t} + \sum_{i=1}^{n} \frac{\partial}{\partial y_i}[a_i(\mathbf{y}, t) p(\mathbf{y}, t|\mathbf{x}, s)] - \frac{1}{2} \sum_{i,j=1}^{n} \frac{\partial^2}{\partial y_i \partial y_j}[b_{ij}(\mathbf{y}, t) p(\mathbf{y}, t|\mathbf{x}, s)] = 0, \tag{2.120}$$

where $\mathbf{x} = (x_1, \ldots, x_n)$, $\mathbf{y} = (y_1, \ldots, y_n)$ denote the states of the process, and $a_i(\mathbf{y}, t)$, $b_{ij}(\mathbf{y}, t)$ are the components of the vector-function $\mathbf{A}(\mathbf{y}, t)$ and the matrix-function $\mathbf{B}(\mathbf{y}, t)$, respectively; like $a(y, t)$ and $b(y, t)$ in the one-dimensional case, the functions $\mathbf{A}(\mathbf{y}, t)$ and $\mathbf{B}(\mathbf{y}, t)$ characterize the infinitesimal properties of the process $\mathbf{X}(t)$.

An important example of a Markov diffusion is the *Brownian motion process* or the *Wiener process* — a mathematical model of the Brownian motion of a free particle immersed in a liquid. Thorough understanding of the Wiener process proves particularly important, since many other diffusion Markov processes can be constructed using the Wiener process. (See [77].)

A stochastic process $X(t)$, $t \in [0, \infty]$, is said to be a *Wiener process* and is denoted by $W(t)$ if it has the following properties:

a) $P\{W(0) = 0\} = 1$.
b) The increments $W(t) - W(s)$, $s<t$ are independent and Gaussian.
c) $\langle W(t) - W(s) \rangle = 0$.
d) $\langle [W(t) - W(s)]^2 \rangle = \sigma^2 (t-s)$, where σ^2 is a positive constant.

The correlation function of a Wiener process is

$$K_W(t_1, t_2) = \sigma^2 \min(t_1, t_2). \tag{2.121}$$

The density of the probability of a transition from state x to state y is given by

$$p(y, t|x, s) = \frac{1}{\sigma\sqrt{2\pi(t-s)}} \exp\left[-\frac{(y-x)^2}{2(t-s)\sigma^2}\right]. \tag{2.122}$$

10.5 Cumulative Jump Processes

An important class of stochastic processes can be described as follows. A process starts its evolution from a state x_0 (of its state space). It stays there a random length of time, say, τ_0, and then jumps to another state x_1, where it remains for a random length of time τ_1, and then jumps to a state $x_2 \neq x_1$, and so on. The evolution repeats this process. The most common representation of such jump processes is a *Poisson process*, for which a state space consists of discrete points: $x = 0, 1, 2, \ldots$ Poisson processes arise in situations where one is interested in the total number of occurrences of a *specified type* of event in the time interval $[0, t]$.

A stochastic process $N(t)$ with independent increments, $t \geq 0$, is said to be a Poisson process with intensity $\lambda_0 > 0$, if $N(0) = 0$ and

$$P\{N(t) = k\} = \frac{(\lambda_0 t)^k}{k!} e^{-\lambda_0 t}, \qquad k = 0, 1, 2, \ldots . \tag{2.123}$$

The mean and variance of $N(t)$ are respectively given by

$$\langle N(t) \rangle = \lambda_0 t, \qquad \sigma_N^2(t) = \lambda_0 t. \tag{2.124}$$

Every process with independent increments, and hence a Poisson process, is a Markov process. It can be shown [55] that if $\{t_n\}$, $t \geq 1$, is the sequence of successive occurrence times in a Poisson process $N(t)$, then the probability density $f_{t_n}(t)$ of the nth occurrence time t_n is the density of the gamma distribution $G(a, p)$ (Eq. (2.28)) with $a = \lambda_0$ and $p = n - 1$, i.e.,

$$f_{t_n}(t) = \lambda_0 e^{-\lambda_0} \frac{(\lambda_0 t)^{n-1}}{(n-1)!}, \qquad t \geq 0. \tag{2.125}$$

In some situations, the intensity of transition from one state to another may depend on time, i.e., $\lambda = \lambda(t)$, $t \geq 0$. This leads to an *inhomogeneous Poisson process*, whose probability distribution is given by

$$P\{N(t) = k\} = \frac{[\lambda(t)]^k}{k!} e^{-\lambda(t)}, \qquad k = 0, 1, 2, \ldots . \tag{2.126}$$

Clearly, an ordinary (or, homogeneous) Poisson process is a special case when $\lambda(t) = \lambda_0 t$.

Another interesting situation occurs when the transition intensity depends on the state of the process (as in population growth, where birth rate depends on the size of the population). In such cases, $\lambda = \lambda(x)$. Discontinuous, or jump, Markov processes with state space $\{0, 1, 2, \ldots\}$ and with $\lambda = \lambda(k)$, $k = 0, 1, 2, \ldots$ are known as *birth processes*. The simplest case is a *linear birth process* in which the intensity, or birth rate, is $\lambda = \lambda^0 k$, where λ^0 is a positive constant. With the assumption that

$P_1(0) = 1$ and $P_k(0) = 0$ for $k \neq 1$, the probability distribution of a (pure) linear birth process is given by

$$P_k(t) = P\{N(t) = k\} = e^{-\lambda^0 t}(1 - e^{-\lambda^0 t})^{k-1}, \quad t > 0, k \geq 1,$$
$$P_0(t) = 0. \tag{2.127}$$

The probability distribution given by Eq. (2.127) is often termed the *Furry–Yule* distribution. It can be shown that the mean and variance of the linear birth process are given by [57]

$$m_N(t) = \langle N(t) \rangle = e^{\lambda^0 t},$$
$$\sigma_N^2(t) = (e^{2\lambda^0 t} - e^{\lambda^0 t}) = e^{\lambda^0 t}(e^{\lambda^0 t} - 1). \tag{2.128}$$

A general class of stochastic jump processes can be obtained if we assume that the intensity is an arbitrary function of time t and state x, and that the state space is continuous, i.e., $x \in R_1$. Hence, a change of a state may occur only by jumps, and if a process is at state x in time t, then the probability that at time $t + \Delta t$ it will be in another state (different then x) is equal to $\lambda(t, x)\Delta t$. For a wide class of such Markov processes, the transition probability density function satisfies a differential integral equation known as the *Feller–Kolmogorov equation.* (See [57].)

An important class of stochastic processes, which can serve as adequate model processes for fatigue, are *cumulative jump processes*, defined as a sum of a random number of random components. In general, these processes can be represented as follows:

$$X(t) = \sum_{k=1}^{N(t)} w(t, \tau_k, Y_k), \quad t \geq 0, \tag{2.129}$$

where $w(t, \tau, y)$ is a suitable function of three real variables; the Y_k are a sequence of random variables independent of $N(t)$; and τ_k represents the time instant in which an event (e.g., occurrence of random shock) takes place. Y_k usually denotes the effect associated with an event occurring at τ_k, and $N(t)$ is an appropriate counting stochastic process, i.e., a process characterizing a number of events occurring in interval $[0, t]$; in particular, it can be assumed to be a Poisson or birth process.

For most practical situations, it is sufficient to consider the subclass of Eq. (2.129), represented as

$$X(t) = \sum_{k=1}^{N(t)} Y_k w(t, \tau_k) . \tag{2.130}$$

In this case, $Y_k w(t, \tau_k)$ characterizes the values at time t of a physical effect due to an event that has occurred at time τ_k. The process $X(t)$ describes the cumulative effect at time t. A special case of Eq. (2.130) is the *cumulative process with exponential decay,*

$$X(t) = \sum_{k=1}^{N(t)} Y_k e^{-\alpha(t-\tau_k)} . \tag{2.131}$$

In the analysis of the cumulative jump processes defined before, it is usually assumed that the random variables Y_k are statistically independent and they have the same probability distribution.

A common special case of the process described by Eq. (2.130) is the so-called *compound Poisson process*, obtained when $w(t, \tau_k) = 1$, i.e.,

$$X(t) = \sum_{k=1}^{N(t)} Y_k, \tag{2.132}$$

where $N(t)$ is a Poisson process, and the Y_k are independent, identically distributed random variables. This process has independent increments and is a Markov process. Its mean and variance are given by

$$\begin{aligned} m_X(t) &= \langle X(t) \rangle = \lambda_0 t \langle Y \rangle, \\ \sigma_X^2(t) &= \lambda_0 t \langle Y^2 \rangle. \end{aligned} \tag{2.133}$$

Cumulative jump models for fatigue crack growth will be discussed in Chapter V.

10.6 Approximations and Simulation

In many situations, the analysis of complicated processes, including random fatigue loads and fatigue accumulation, can only be performed when the process in question is properly approximated. In addition, digital simulation of the random processes often appears to be the only feasible method of analysis. Approximation of stochastic processes and their simulation is a rather vast field, and while this is not the proper place to discuss these problems in detail, it seems worthwhile to at least make several remarks.

One of the most common methods for approximating functions in deterministic analysis is polynomial approximation — grounded in the

Weierstrass theorem. However, Weierstrass-type theorems can also be formulated for stochastic processes and random functions of several variables. Consider the random polynomial of degree n defined by

$$P_n(t, \gamma) = \sum_{k=0}^{n} a_k(\gamma) t^k, \tag{2.134}$$

where $a_k(\gamma)$ are random variables. A possible extension of the Weierstrass approximation theorem to random processes is given in the next few paragraphs. (See [70].)

If $X(t, \gamma)$ is a random process, continuous in probability in the interval $I \subset R_1$, then there exists a family of random polynomials $\{P_n(t, \gamma)\}$ converging uniformly in probability to $X(t, \gamma)$; i.e., for each $\epsilon > 0$, $\eta > 0$, there exists an integer $N = N(\epsilon, \eta)$ such that, for all $n \geq N$ and any $t \in I$,

$$P\{\gamma: |X(t, \gamma) - P_\eta(t, \gamma)| \geq \epsilon\} \leq \eta. \tag{2.135}$$

One can also approximate a general stationary processes (with continuous spectrum) by *discrete* trigonometric polynomials.

Theoretically, we have the following assertion: For an arbitrary stationary process $X(t, \gamma)$ and for any $\epsilon > 0$ and arbitrary T (sufficiently large), there exist pairwise uncorrelated random variables $a_1(\gamma), \ldots, a_n(\gamma)$ and $b_1(\gamma), \ldots, b_n(\gamma)$ and real numbers $\omega_1, \ldots, \omega_n$ such that for an arbitrary $t \in [-T, T]$,

$$\left\langle \left[X(t, \gamma) - \sum_{k=1}^{n} (a_k(\gamma) \cos \omega_k t + b_k(\gamma) \sin \omega_k t) \right]^2 \right\rangle \leq \epsilon. \tag{2.136}$$

For simulation purposes, we would like to have a procedure for calculating ω_k, a_k, and b_k. In this context, the work of Pakula [72] is of interest. However, the construction of the preceding quantities provided in this paper is rather complicated.

Practical digital simulations are most often accomplished by use of a finite sum of cosine functions with random phase angles of the form (see Shinozuka [73], [74])

$$X(t, \gamma) = \sqrt{2} \sum_{k=1}^{N} A_k \cos[\omega_k t - \psi_k(\gamma)], \tag{2.137}$$

where the ω_k are realized values of frequency, distributed according to the spectral density $g_X(\omega)$, and the $\psi_k(\gamma)$ are mutually independent random variables distributed uniformly on the interval $[0, 2\pi]$, i.e.,

$$A_k = [\tilde{g}_X(\omega_k)\,\Delta\omega]^{1/2},$$
$$\omega_k = (k - \frac{1}{2})\,\Delta\omega, \qquad \tilde{g}_X(\omega) = 2g_X(\omega). \tag{2.138}$$

The accuracy of the approximation in representing the actual process depends on whether the interval of frequencies $[0, \omega^*]$ taken into account in discretization is large enough and whether $\Delta\omega$ is small enough such that the relation

$$A_k^2 = \int_{(k-1)\Delta\omega}^{k\Delta\omega} \tilde{g}_X(\omega)\,d\omega \approx \tilde{g}_X(\omega_k)\,\Delta\omega, \tag{2.139}$$

is valid. It is clear that the larger the value of N used in the simulation, the better the accuracy of the simulated process.

In the approach to simulation just described, the process under consideration is represented by a finite number of cosine functions and the stochastic aspects are reduced to the selection of appropriate amplitudes and phases. The process itself is discrete in the frequency domain and continuous in time. All frequency components contribute at any given time, and the amount of computation in each time step grows rapidly with the required frequency band. Although the use of the fast Fourier transform algorithm improves the situation, this requires that the full time history be calculated simultaneously.

In recent years, sequential simulation algorithms have been introduced in the form of Moving-Average (MA), Auto-Regressive (AR), and the combination Auto-Regressive Moving-Average processes (ARMA). These processes are discrete in time and continuous in the frequency domain. The general form of the ARMA model is (e.g., see Kozin [67]; Krenk and Clausen [68])

$$X_n + \sum_{k=1}^{N} a_k X_{n-k} = \sum_{k=0}^{M} b_k \xi_{n-k}, \tag{2.140}$$

where $\{a_k\}$, $\{b_k\}$ are constant coefficients, and $\xi_n, \xi_{n-1}, \ldots, \xi_{n-M}$ is a sequence of independent, identically distributed random variables (most often taken to be Gaussian variables). The sequence $\{X_n, X_{n-1}, \ldots\}$ denotes the discrete values (observations at discrete instants of time) of the process to be simulated and is called a *time series*.

The model given in Eq. (2.140) is of order (N, M), and all variables are scalar quantities. This corresponds to a so-called single-input–single-output linear model. ARMA models can be extended to multi-input and multi-output linear systems, in which case the observations X_k and the random terms ξ_k become vectors, and the coefficients $\{a_k\}$, $\{b_k\}$ become matrices.

As is seen from Eq. (2.140), ARMA models correspond to passing a white noise through a discrete filter. Available methods of analysis for digital filters can be used to describe the properties of the simulated sequence (e.g., Oppenheim and Schafer [71]).

The MA process corresponds to $N = 0$ in Eq. (2.140), and in this case, the value of the process at time $t = n$ is characterized by a linear combination of Gaussian random variables. The AR process corresponds to $M = 0$; this means that X_n is generated from a linear combination of N previous values of the process plus a single independent random variable.

The basic problem of modelling a time series $\{X_n\}$ by the ARMA process consists of estimating the parameters $\{a_k\}$, $\{b_k\}$. There is also the question of determining the best model order to fit the observed data (i.e., the best choice of N and M; see [67]).

The ARMA process has many advantages and has received considerable attention in a variety of disciplines (e.g., simulation of load processes such as wind, sea waves, and earthquakes). It is worth noting that the AR representation is equivalent to a model of a time series that has maximum entropy. Furthermore, the ARMA process turned out to be particularly suited for random vibration analysis of linear systems; it can be shown that the ARMA process with $N = M + 1 = 2n$ is an exact representation of the discretely sampled response of an n-degree of freedom linear system subjected to a white noise excitation. (See Gersch and Liu [64].)

Chapter III

Random Fatigue Loads: Statistical Characteristics

11. Introductory Remarks

Throughout their service life, machines, vehicles, and buildings are subjected to loads, the majority of which vary with time in a very complicated manner. These loadings may originate from external excitations such as wind gusts, sea waves, noise, or road roughness, but they may also result from the usage of the structure (e.g., from loading or unloading a container or accelerating a vehicle). In general, all these *factors* result in a complicated stress distribution within a material that directly affects the fatigue process.

Most traditional fatigue analyses are based upon a representation of loads in the form of periodic deterministic functions of time, and basic characteristics of fatigue accumulation are expressed in terms of the number of loading cycles. At present, it is widely accepted that fatigue analyses performed under constant amplitude or block loading insufficiently reflect the complexity of the fatigue process under actual loadings. Irregular time histories for the load — including random loads — have to be considered and suitably modelled.

Fatigue accumulation is mainly due to a sequence of peaks/troughs of a random stress process, or to stress reversals. This observation is reflected in the crack growth equations for deterministic stress conditions in which the basic quantities are the stress range $\Delta S = S_{max} - S_{min}$ and the maximum stress values S_{max}. (See Section 4.2.) In dealing with random loadings, it is therefore important to characterize its extremal values. The quantities that we have in mind are, for example, the time spent by a process above a fixed level u, the number of times a trajectory crosses the level u in a time interval of duration T, the number and distribution of maxima in time T, etc. Recent mathematical theory of the problems indicated is presented in the book by Leadbetter et al. [69]. (See also Cramer and Leadbetter [61].) In this chapter, we will provide a brief synopsis of the most useful results.

12. Nature of Fatigue Loads

Depending upon the type of engineering structure and its operational task, we meet various kinds of load processes. Fatigue loads experienced by bridges, cranes, and pressure vessels are commonly perceived as discrete events in time and are often modelled by sequences of discrete variable-amplitude loads occurring randomly in time. However, aircraft structures are, in general, subjected to excitations that fluctuate in time continuously.

Structures exposed to gusty wind (e.g., tall buildings and transport aircraft) experience a fluctuating wind pressure created by turbulence. The magnitude and character of the load fluctuations depend on the shape of the structure and its orientation with respect to the wind direction. Wind loading can be derived from statistical data for wind velocity. Relatively little data exist for direct wind forces. A complete description of wind forces requires consideration of the variation of the wind velocities from point to point on the structure, as well as the response of the structure itself. (See [87], [106].) To convert instantaneous wind velocity $V(t)$ to wind pressure $P(t)$ acting on a particular part of a structure, the standard hydrodynamic relationship is employed, i.e.,

$$P(t) = \frac{1}{2}\rho C V^2(t), \tag{3.1}$$

where ρ is the density of air (about 1.2 kg/m^3) and C is the *wind pressure coefficient*, which depends on the size and orientation of the structure.

A convenient approach is to consider the wind velocity $V(t)$ to be composed of a time-independent mean value v_0 and an additive stationary stochastic process $V'(t)$, assumed to be much smaller than v_0. In such a case, Eq. (3.1) is usually taken in the form,

$$P(t) = \frac{1}{2}\rho C\,[v_0^2 + 2v_0 V'(t)], \tag{3.2}$$

where $V'(t)^2$ has been neglected. If there is more than one wind velocity component, v_0 and $V'(t)$ should be replaced by vectors. When the structure also responds with its own velocity, then $V'(t)$ should be replaced by the relative velocity.

The probabilistic characteristics of turbulent wind fluctuations depend on meteorological conditions, the geographical position, the height over the Earth's surface, etc. Statistical inference of these characteristics is not straightforward. Without going into details, we wish to say that the engineering analysis of fatigue reliability assumes some *standard* representations of the spectrum of a turbulent wind. For example, in the analysis of aircraft structures, the following form for the spectral density of the longitudinal component of stationary random velocity is often assumed [82]:

$$g_V(\omega) = \frac{\sigma_Z^2 \lambda_V}{\pi v_0} \frac{1 + 3\tilde{\omega}^2}{(1 + \tilde{\omega})^2}, \qquad \tilde{\omega} = \frac{\lambda_V}{v_0}\omega, \tag{3.3}$$

where σ_Z^2 is the variance of the vertical component, and λ_V denotes the scale of the turbulence in the longitudinal direction.

Another important type of structure subjected to significant fatigue loads is the steel offshore platform. Fatigue loading acting on offshore structures is generated by wind and sea waves. Waves, in turn, occur as a result of complicated interaction between wind and water. This leads to a loading process that is often described by a series of continuously varying sea states. The nature of offshore loading and the complex interactions likely in the seawater environment make establishment of standard load spectra for offshore structures much more difficult than for aircraft structures.

Although sea motion (or sea states) can be partially characterized by some parameters (e.g., the wave height h_s, the mean wave period T_s, the wave direction θ), an underlying quantity in stochastic theory is the sea elevation $\eta(x, y, t)$, which is regarded as a random function of position and time. Probabilistic properties of $\eta(x, y, t)$ are derived partially from the measurements and partially from hydrodynamic wave theory. In almost all studies in ocean engineering, it is assumed that the sea wave process is a stationary stochastic process. Under such a hypothesis, the process $\eta(x, y, t)$, for fixed (x, y), is characterized by the spectral density $g_\eta(\omega)$.

Various forms of the spectral density of sea surface elevation $\eta(t)$, for fixed (x, y), have been proposed in the literature. The most popular form used in practice is the Pierson–Moskowitz spectrum,

$$g_\eta(\omega) = \frac{A a_g^2}{\omega^5} \exp\left[-B\left(\frac{a_g}{\omega v_0}\right)^4\right], \qquad \omega > 0, \tag{3.4}$$

where A and B are dimensionless constants taken to be $A = 8.1 \times 10^{-3}$, $B = 0.74$; v_0 is the mean wind velocity at a height of 19.5 m above the still sea surface; and a_g is the acceleration of gravity. Using a linearized wave theory, one obtains a relationship between the fluid particle velocity $u(x, y, z, t)$ and the surface elevation $\eta(x, y, t)$, and, as a consequence, an expression for spectral density of the fluid particle velocity. (See [31].)

Once the water particle velocity and acceleration are known as a function of the depth z, the normal force per unit length exerted at a particular location on a slender cylinder (such as is typical in steel offshore structures) is given by the Morison equation,

$$\mathbf{F}(t) = k_d \mathbf{u} | \mathbf{u} | + k_m \dot{\mathbf{u}}, \tag{3.5}$$

where $\mathbf{u}$ is the vector of incident water particle velocity normal to the cylinder. The first term in Eq. (3.5) relates to the drag force exerted by the wave on the pile, and the second term relates to oscillatory behavior. One should notice that the force $\mathbf{F}(t)$ is nonlinearly related to the fluid particle velocity as well as the sea elevation. Thus, even if η and $\mathbf{u}$ would be Gaussian processes, the loading process $\mathbf{F}(t)$ will not be Gaussian. Although it is often adopted in engineering practice, the nonlinearities in Eq. (3.5) invalidate the assumption that the sea elevations follow a Gaussian distribution. (See Ochi [97].) The non-Gaussian character of the forces acting on offshore structures causes additional problems with their proper characterization. The spectral density, in this situation, provides only a partial characterization of the process. Higher-order statistics, as well as polyspectra, should be estimated from the data.

The preceding description of two types of random loading (i.e., turbulent wind loading of aircraft structures and sea wave loading of offshore structures) indicates the methods and potential difficulties in characterizing fatigue loads experienced by structures. One way to describe a random fatigue load is to report the number of peaks and troughs at each stress level; this may be done either by actual load measurements or by appropriate simulation techniques. Another way of specifying random fatigue loads is through a limited number of statistics that reflect, on the average, the nature of the loading process. Such statistics — especially relevant to fatigue — will be presented in the next section. A reader interested in the details of practical characterization of various fatigue loads is referred to the publication edited by Potter and Watanabe [99], as well as to Albelkis and Potter [81].

13. Extremal Statistics of Gaussian Loads

13.1 Fractional Occupation Time

Let $S(t)$ be a real-valued stationary stochastic process with continuous trajectories on the considered interval $[0, T]$, and with zero mean, correlation function $K_S(\tau)$, and spectral density $g_S(\omega)$.

The fractional occupation time l_u is understood as the proportion of time $0 \le t \le T$ that the process $S(t)$ spends above the level u. (See Fig. 3.1.) If we introduce a zero–one process defined as

$$\eta(t;u) = \begin{cases} 1, & \text{if } S(t) > u \\ 0, & \text{if } S(t) \le u, \end{cases} \tag{3.6}$$

then the random variable l_u is represented as

$$l_u = \frac{1}{T}\int_0^T \eta(t;u)\,dt. \tag{3.7}$$

The mean value of l_u is

$$\langle l_u \rangle = \frac{1}{T}\int_0^T P\{X(t) > u\}\,dt. \tag{3.8}$$

If process $S(t)$ is stationary and Gaussian, then

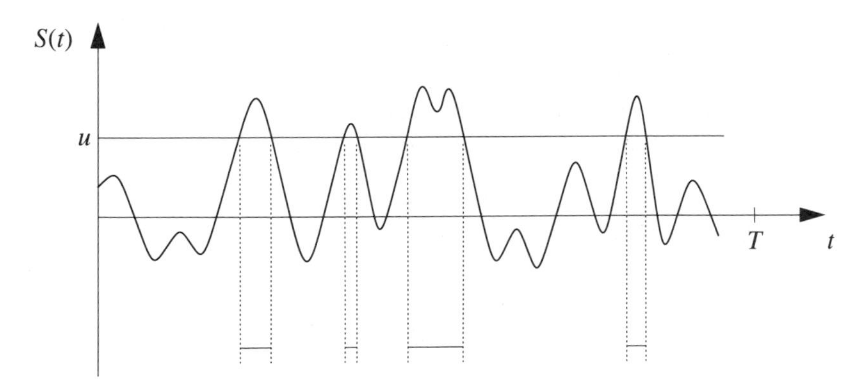

Figure 3.1. Sample function of S(t) and fractional occupation time.

$$\langle l_u \rangle = 1 - \Phi\left(\frac{u}{\sqrt{K_S(0)}}\right), \quad K_S(0) = \sigma_S^2 = \lambda_0, \tag{3.9}$$

where $\Phi(x)$ is the standard normal integral (or Laplace function) given by

$$\Phi(x) = \frac{1}{\sqrt{2\pi}} \int_{-\infty}^{x} e^{-\xi^2/2} d\xi. \tag{3.10}$$

The variance of l_u is as follows:

$$\sigma_{l_u}^2 = \frac{1}{T^2} \sum_{\nu=1}^{\infty} \frac{1}{\nu!} \left[\Phi^{(\nu)}\left(\frac{u}{\sigma_S}\right)\right]^2 \int_0^T \int_0^T \rho^{\nu}(t_2 - t_1)\, dt_1 dt_2, \tag{3.11}$$

where $\rho(t_2 - t_1) = K_S(t_2 - t_1)/\sigma_S$, and $\Phi^{(\nu)}$ denotes the νth derivative of the function Φ given by Eq. (3.10). If $K_S(\tau) = \sigma_S^2 \exp(-\alpha|\tau|)$, for example, then

$$\sigma_{l_u}^2 = \frac{2}{\alpha T} \sum_{\nu=1}^{\infty} \frac{1}{\nu!\,\nu} \left[\Phi^{(\nu)}\left(\frac{u}{\sigma_S}\right)\right]^2 \left[1 + \frac{1}{\nu\alpha T}\left(e^{-\nu\alpha T} - 1\right)\right]. \tag{3.12}$$

It is seen that if the length of observation interval T is sufficiently large and the correlation between the value of the process decays quickly, then the contribution of the factor $[1 + (e^{-\nu\alpha T} - 1)/\nu\alpha T]$ is small and can be neglected. Most often, only a few terms of the series in Eqs. (3.11) and (3.12) contribute significantly to the sum.

13.2 Level Crossings

Let $S(t)$ be a stationary and Gaussian process and let $N_u(0, T)$ denote the number of crossings of the level u by the trajectories of the stochastic process $S(t)$ within a given interval $[0, T]$. (See Fig. 3.2.) The average number of crossings of the level u within interval $[0, T]$ is

$$\begin{aligned} \langle N_u(0, T) \rangle &= T\frac{1}{\pi}\left(\frac{K_{\dot{S}}(0)}{K_S(0)}\right)^{1/2} \exp\left(-\frac{u^2}{2K_S(0)}\right) \\ &= T\frac{1}{\pi}\left(\frac{\lambda_2}{\lambda_0}\right)^{1/2} \exp\left(-\frac{u^2}{2\lambda_0}\right), \end{aligned} \tag{3.13}$$

where λ_0 and λ_2 are the spectral moments defined by Eq. (2.104), and

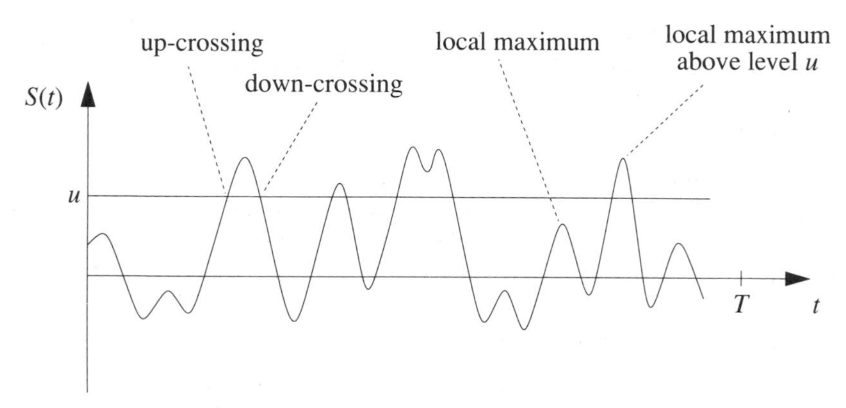

Figure 3.2. Sample function (trajectory) of $S(t)$, its crossing of a level u and extremes.

$$K_{\dot{S}}(0) = -\left.\frac{d^2 K_S(\tau)}{d\tau^2}\right|_{\tau=0} \tag{3.14}$$

is assumed to be finite. If $K''_S(0)$ is infinite, then the right-hand side of Eq. (3.13) is infinite. (For the extension to nondifferentiable processes; see [69].) Eq. (3.13) is commonly known as *Rice's formula.*

The expected number of zero-crossings $(u=0)$ has the simple form,

$$\langle N_0(0, T) \rangle = T\,\frac{1}{\pi}\left(\frac{\lambda_2}{\lambda_0}\right)^{1/2}. \tag{3.15}$$

For a narrow-band process where the spectral density $g_s(\omega)$ has significant values only in a frequency band whose width is small compared with the value of the mid-band frequency, each cycle corresponds to a zero-crossing with positive or negative slope. Hence, for a stationary Gaussian and narrow-band process, the expected number of zero-crossings, say, with positive slope is equal to a number of *equivalent* cycles per unit time, say, n_e. So, the *equivalent frequency* ω_e can be defined as $n_e = \omega_e/2\pi$. Therefore,

$$\omega_e = 2\pi\langle N_0^+(0, 1) \rangle = \left(\frac{\lambda_2}{\lambda_0}\right)^{1/2} = \frac{\sigma_{\dot{S}}}{\sigma_S}. \tag{3.16}$$

The problem of determining the probability distribution of the number of crossings of level u is rather complicated. The existing results are only concerned with some limiting distributions as $T \to \infty$ or $u \to \infty$.

Remark 1. If the process $S(t)$ has a nonzero mean m_S, it is only necessary to replace u by $u - m_S$ in the preceding results, since $S(t)$ crosses the level u when $S(t) - m_S$ crosses the level $u - m_S$. Also, we have $\langle N_u(0, T)\rangle = T\langle N_u(0, 1)\rangle$.

Remark 2. The results concerned with the crossing of a given level can be generalized to the case where a *level* is not a horizontal line, but a differentiable function $u(t)$. Such a problem can be regarded as a zero-crossing problem for a nonstationary process $X(t) = S(t) - u(t)$.

13.3 Local Maxima (Peaks)

There are two basic problems associated with investigation of local maxima (or peaks) of a given stochastic process. The first one consists of characterizing the number of maxima within a given time interval. The second problem is associated with determination of the distribution function for the height of a local maximum, given that such a maximum occurs.

Let $M_u(t_1, t_2)$ denotes the *number of maxima* of $S(t)$ within the time interval $[t_1, t_2]$ above the level u. If process $S(t)$ is standardized (i.e., zero mean and unit variance), stationary, Gaussian, and twice differentiable (i.e., the fourth spectral moment λ_4 should be finite), then the average number of maxima $M_u(0, T)$ is

$$\langle M_u(0, T)\rangle = T\,\frac{1}{2\pi}\left\{\left(\frac{\lambda_4}{\lambda_2}\right)^{1/2}\left[1 - \Phi\left(u\sqrt{\frac{\lambda_4}{\Delta}}\right)\right] + (2\pi\lambda_2)^{1/2}\varphi(u)\,\Phi\left(\frac{u\lambda_2}{\sqrt{\Delta}}\right)\right\}, \tag{3.17}$$

where $\varphi(u)$ is the probability density of the standard normal distribution function and $\Delta = \lambda_4 - \lambda_2^2$. It is seen that the expected *total* number of maxima as $u \to -\infty$ is

$$\langle M_{-\infty}(0, T)\rangle = T\,\frac{1}{2\pi}\left(\frac{\lambda_4}{\lambda_2}\right)^{1/2}. \tag{3.18}$$

To obtain relations for direct application to the case where the variance $\sigma_S^2 = \lambda_0 \neq 1$, a slight change in Eq. (3.17) is needed, since $\underset{\sim}{S}(t) = S(t)/\sigma_S$ satisfies $\sigma_{\underset{\sim}{S}}^2 = 1$. Introducing the so-called *spectral width parameter*,

$$\epsilon = (1-\alpha^2)^{1/2}, \quad \alpha = \frac{\lambda_2}{\sqrt{\lambda_0 \lambda_4}}, \tag{3.19}$$

where α, the regularity factor, is defined according to Eq. (2.105), the average number of local maxima above level u in $[0, T]$ of the process $S(t)$, which is stationary and Gaussian with zero mean and variance λ_0, is

$$\begin{aligned}&\langle M_u(0,T)\rangle \\ &= T\left\{\nu_2\left[1-\Phi\left(\frac{u}{\epsilon\sqrt{\lambda_0}}\right)\right]+\nu_0\exp\left(-\frac{u^2}{2\lambda_0}\right)\Phi\left(\frac{u}{\epsilon\sqrt{\lambda_0}}\frac{\nu_0}{\nu_2}\right)\right\},\end{aligned} \tag{3.20}$$

where

$$\nu_0 = \frac{1}{2\pi}\left(\frac{\lambda_2}{\lambda_0}\right)^{1/2}, \quad \nu_2 = \frac{1}{2\pi}\left(\frac{\lambda_4}{\lambda_2}\right)^{1/2}. \tag{3.21}$$

Therefore, the average number of maxima for a stationary and Gaussian process is entirely characterized by the spectral moments λ_k and the spectral width parameter ϵ. One should notice that the regularity factor α is the ratio of the average number of zero-crossings to the expected number of local maxima. Its value lies between 0 and 1. A narrow-band random process has (approximately) an equal number of peaks and zero-crossings with positive (or negative) slope, so $\alpha \to 1$; this means that for a narrow-band process, $\epsilon \to 0$, and

$$\langle M_u(0,T)\rangle = T\,\frac{1}{2\pi}\left(\frac{\lambda_2}{\lambda_0}\right)^{1/2}\exp\left(-\frac{u^2}{2\lambda_0}\right). \tag{3.22}$$

Let us denote the probability density of the height Z of the peaks by $f_{max}(z)$. Under the assumptions made earlier (i.e., $S(t)$ is stationary and Gaussian with zero mean and variance $\sigma_S^2 = \lambda_0$), the following result is valid [69],

$$\begin{aligned} f_{max}(z) &= \frac{\epsilon}{\sqrt{\lambda_0}}\varphi\left(\frac{z}{\epsilon\sqrt{\lambda_0}}\right) \\ &+ \sqrt{1-\epsilon^2}\,\frac{z}{\lambda_0}\exp\left(-\frac{z^2}{2\lambda_0}\right)\Phi\left(\frac{z(1-\epsilon^2)^{1/2}}{\epsilon\sqrt{\lambda_0}}\right),\end{aligned} \tag{3.23}$$

where ϵ is defined by Eq. (3.19), and φ and Φ are the standard normal density and distribution functions, respectively. For a *narrow-band process* (i.e., when $\epsilon \to 0$), the height of peak magnitudes has a Rayleigh distribution,

$$f_{\max}(z) = \frac{z}{\sigma_S^2} \exp\left(-\frac{z^2}{2\sigma_S^2}\right), \quad z \geq 0. \tag{3.24}$$

This distribution is of special practical interest, because the stationary response of a lightly damped single degree of freedom linear system to a stationary Gaussian excitation can be considered as a narrow-band Gaussian process.

On the other hand, when α is very small, i.e., $\epsilon \to 1$, we have a situation characteristic of a broad-band random process where the average number of peaks is much larger than the expected number of zero-crossings. Therefore, for a *broad-band* process (when $\epsilon \to 1$), the distribution (3.23) reduces to

$$f_{\max}(z) = \frac{1}{\sigma_S \sqrt{2\pi}} \exp\left(-\frac{z^2}{2\sigma_S^2}\right), \tag{3.25}$$

which is a Gaussian distribution.

For random loads that can be considered as narrow-band processes, the corresponding peaks and troughs can be assumed to occur at levels $S_{\max} = m_S + Z$ and $S_{\min} = m_S - Z$, where Z is the random height of peaks with distribution (3.24). Hence, the range $\Delta S = H = S_{\max} - S_{\min}$ is equal to $2Z$. Estimation of fatigue accumulation, whether related to peaks of ranges, may thus be related to the distribution (3.24).

Remark 3. As the load bandwidth increases, narrow-band models ignore two important effects: (i) The number of small-amplitude, high-frequency oscillations grows, so that actual peak and range values are less (on average) than the narrow-band model predicts; and (ii) life estimation from constant-amplitude results requires identification of the larger-amplitude, lower-frequency cycles formed by the multiple peaks and troughs that are very damaging. The first effect leads to conservative errors in the narrow-band model, while the second effect introduces nonconservatism. Usually, the first effect tends to dominate so that the narrow-band models are generally regarded to be conservative.

13.4 Range (Wave Height)

The results given previously have been concerned with a single local extremum. A much more difficult problem is to relate one local extremum to the next. There are two quantities of great practical significance (especially in oceanography and fatigue analysis): (i) the time interval between consecutive local extrema (wavelength); and (ii) their difference in magnitude (range or wave height H).

In analysis of sea waves in oceanography, the severity of the seas is often described in terms of the *significant wave height* or the *significant amplitude*. Let $z_{1/3}$ be such that

$$P(A \geq z_{1/3}) = 1 - F_{\max}(z_{1/3}) = \frac{1}{3}, \tag{3.26}$$

that is, one-third of the amplitudes exceeds $z_{1/3}$, and define

$$A_S = 3 \int_{z_{1/3}}^{\infty} z\, f_{\max}(z)\, dz \tag{3.27}$$

to be the mean of the one-third highest waves. Then A_S is called the significant amplitude. The significant wave height is defined similarly, and one often assumes that $H_S = 2A_S$. This approximation is better for narrow-band processes when ϵ is small. For small values of ϵ, the following approximation of the significant amplitude A_S of stationary and Gaussian processes is often used:

$$\begin{aligned} A_S &\approx \sqrt{\lambda_0}\,\{\sqrt{2\ln 3} + 3\sqrt{2\pi}\,[1 - \Phi(\sqrt{2\ln 3})]\,\} \\ &\approx 2.00\sqrt{\lambda_0} = 2.00\,\sigma_S\,. \end{aligned} \tag{3.28}$$

The preceding characterization of random sea waves can also be useful in modelling and analysis of fatigue (especially due to sea wave pressure on offshore structures).

Many attempts have been made to find the exact form of the distribution of wavelength and amplitude of random processes, but no theoretically founded closed-form expression has been found. However, some approximations and suggestions exist. (See Lindgren and Rychlik [92], Rice and Beer [101], Chakrabarti and Cooley [83], Lindgren [91] and Longuet–Higgins [94].)

Closing the remarks on the distribution of ranges, it is worth adding that using a sinusoidal approximation of the trajectory of a random process between adjacent extrema, Ortiz [98] has derived an approximate distribution for the ranges (rises or falls), which is a Rayleigh distribution with parameter $2\alpha\sigma_S$, i.e.,

$$f_R(h) = \frac{h}{4\alpha^2\sigma_S^2}\exp\left(-\frac{1}{2}\frac{h^2}{4\alpha^2\sigma_S^2}\right). \tag{3.29}$$

13.5 Envelope

The behavior of a stationary stochastic process — especially a narrow-band process — resembles a nonrandom function $s(t)$ that oscillates about the mean level. For such oscillatory functions, one can find a nonnegative function $E(t)$, such that $|s(t)| \le E(t)$ for all t, and $|s(t)| = E(t)$ for some t. A function $E(t)$ of this type is known as an *envelope* of the *signal* $s(t)$. In the simple case when $s(t) = a\cos(\omega t + \varphi)$, it is clear that $s(t)$ oscillates between the limits $\pm a$, so it is reasonable to define an envelope of $s(t)$ as $E(t) = a$.

In more complex cases, including stochastic processes, it is not immediately obvious how the envelope should be defined. The earliest definition of an envelope of a random signal (stationary and narrow-band; see Fig. 3.3) is due to Rice [100]. Other definitions have been proposed by Crandall and Mark [84], Cramer and Leadbetter [61], and Krenk [88]; also see [89].

Let $S(t)$ be a real stationary Gaussian process with zero mean. Further, assume that it is a narrow-band process (as originally assumed by Rice). Such a process can be represented as

$$S(t) = E(t)\cos[\omega_0 t + \Phi(t)], \tag{3.30}$$

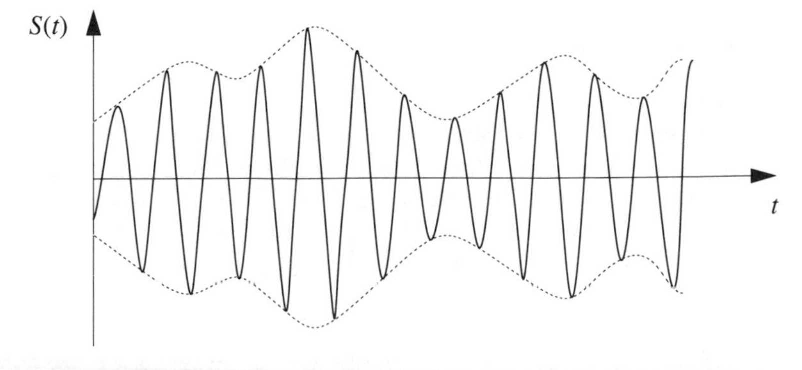

Figure 3.3. A narrow-band stationary process $S(t)$ and its envelope.

where ω_0 is the representative midband frequency of the narrow-band (see Fig. 3.3), and $E(t)$ and $\Phi(t)$ are slowly varying random processes interpreted as the amplitude and phase, respectively. Representation (3.30) can be written alternatively as

$$S(t) = I_c(t)\cos\omega_0 t - I_s(t)\sin\omega_0 t, \tag{3.31}$$

where $I_c(t)$ and $I_s(t)$ are independent Gaussian processes, and

$$\begin{aligned} I_c(t) &= E(t)\cos\Phi(t), \\ I_s(t) &= E(t)\sin\Phi(t). \end{aligned} \tag{3.32}$$

We can interpret $I_c(t)$ and $I_s(t)$ as the coordinates of a random point in the Euclidean plane. It can be shown that $E(t)$ and $\Phi(t)$ are, for each t, independent, and that they have the following distributions:

$$\begin{aligned} f_E(\epsilon) &= \frac{\epsilon}{\sigma_S^2}\exp\left(-\frac{\epsilon^2}{2\sigma_S^2}\right), \quad \epsilon \geq 0, \\ f_\Phi(\varphi) &= \frac{1}{2\pi}, \quad 0 \leq \varphi \leq 2\pi. \end{aligned} \tag{3.33}$$

It is clear that

$$E(t) = [I_c^2(t) + I_s^2(t)]^{1/2}. \tag{3.34}$$

The process $E(t)$ is called the envelope of $S(t)$. Sample functions of $E(t)$ play the role of envelopes of sample functions of $S(t)$; of course, $|S(t)| \leq E(t)$. It should be noted that the one-dimensional distribution (for fixed t) of the envelope $E(t)$ of a narrow-band process is a Rayleigh distribution and coincides with the distribution of peak magnitudes of the process.

If process $S(t)$ is non-Gaussian, the distribution of $I_c(t)$ and $I_s(t)$ is generally different from that of $S(t)$. To find the probabilistic characteristics of the envelope, the joint probability distribution of $I_c(t)$ and $I_s(t)$ should be determined. To avoid such analysis, Crandall and Mark introduced the notion of envelope based on the energy of the randomly oscillating process.

Let us assume that process $S(t)$ has a finite second spectral moment λ_2. Let N_u denote the number of crossings of the level u by $E(t)$ in unit time interval. Then (see Cramer and Leadbetter [61])

$$\langle N_u \rangle = \left(\frac{2\Delta}{\pi}\right)^{1/2} u e^{-u^2/2}, \tag{3.35}$$

where $\Delta = \lambda_2 - \lambda_1^2$.

14. Non-Gaussian Loads

There are situations where a load acting on structure cannot be assumed to be Gaussian. As we have already indicated, this is the case of sea wave loading acting on offshore platforms. An important problem that arises is the effective characterization of such random processes. For the purpose of fatigue analysis, we are primarily interested in the description of extremal properties of non-Gaussian load (stress) processes.

Theoretically, many of the results presented in the previous section are extendable to non-Gaussian processes. For example, Eq. (3.8), for the average value of the time that the process $S(t)$ spends above the level u, is valid in general, and $\langle l_u \rangle$ can be determined when the one-dimensional probability distribution of the process $S(t)$ is given.

Similarly, under appropriate conditions on regularity of the sample functions of the process $S(t)$, the general formula for the average number of crossings of a level u and for the average number of maxima above the level u can be derived. The average number of crossings of the level u is given by

$$\langle N_u(t_1, t_2) \rangle = \int_{t_1}^{t_2} \int_{-\infty}^{\infty} |\dot{s}| f(s, \dot{s}; t)\, d\dot{s}\, dt, \tag{3.36}$$

where $f(s, \dot{s}; t)$ is the joint probability density function of the process $S(t)$ and its derivative $\dot{S}(t)$ at time t. The average number of maxima above the level u is given by

$$\langle M_u(t_1, t_2) \rangle = \int_{t_1}^{t_2} dt \int_{-\infty}^{u} \int_{-\infty}^{0} |\ddot{s}| f(s, 0, \ddot{s}; t)\, d\ddot{s}\, ds, \tag{3.37}$$

where $f(s, \dot{s}, \ddot{s}; t)$ is the joint probability density of $S(t)$, $\dot{S}(t)$, and $\ddot{S}(t)$ at time t.

It is seen that the information about the expected number of crossings and maxima is contained in the joint density functions $f(s, \dot{s}; t)$ and $f(s, \dot{s}, \ddot{s}; t)$, respectively. In practice, however, these functions are not usually available. So, the only approach that will be effective is to approximate $f(s, \dot{s}; t)$ and $f(s, \dot{s}, \ddot{s}; t)$ in terms of Gaussian densities.

A wide class of non-Gaussian processes can be generated by nonlinear transformations of Gaussian processes. To illustrate, let a non-Gaussian process $S(t)$ be represented by

$$S(t) = g(S_G(t)), \tag{3.38}$$

where $S_G(t)$ is a Gaussian process and g is a specified deterministic function. If the function g is a one-to-one mapping, then the probability distributions of $S(t)$ and its derivatives can, in principle, be evaluated. If we assume that g is a monotonic odd function, then the average number of zero-crossings of $S(t)$ and $S_G(t)$ are equal. If, for example, $e_G(t)$ is a sample from the stress envelope $E_G(t)$ of Gaussian process $S_G(t)$, then $g(e_G(t))$ represents the corresponding envelope of $S(t) = g(S_G(t))$.

Lutes et al. [95] considered the case in which

$$S(t) = g(S_G(t)) = aS_G(t)\left[1 + \frac{wS_G^2(t)}{\sigma_{S_G}^2}\right], \tag{3.39}$$

where w is an independent parameter, and a is a constant chosen so that $\sigma_S^2 = \sigma_{S_G}^2$, i.e.,

$$a = (1 + 6w + 15w^2)^{-1/2}. \tag{3.40}$$

The excess coefficient (Eq. (2.14)) is a convenient measure of the departure of a process from Gaussian behavior. The excess coefficient for the process $S(t)$ defined in Eq. (3.39) is given by

$$\zeta_S^{(2)} = 3a^4[1 + 20w + 210w^2 + 126w^3 + 3465w^4] - 3. \tag{3.41}$$

It should be noted that the excess coefficient (and kurtosis) of $S(t)$ increases with w from $\zeta_S^{(2)} = 0$ for $w = 0$ to $\zeta_S^{(2)} = 43.2$ as w goes to infinity.

As has already been mentioned, the non-Gaussian character of the loading of offshore structures is primarily due to the nonlinear relationship (i.e., the Morison relation (3.5)) between the water particle velocity and the wave forces acting on the submerged cylindrical structural element. The mean up-crossing rate of such a non-Gaussian loading process, and its extremal properties, were investigated by Tung [107] and Naess [96]. (See also Winterstein [108].)

15. Nonstationary Loads

Though stationarity is usually assumed in the existing analyses of fatigue under random loading, there are situations where nonstationarity has to be considered. For example, in the analysis of road vehicles, it is usually assumed that: (i) The road roughness is a statistically homogeneous (stationary) random function, and (ii) the traversal velocity of the vehicle is constant. There is no evidence to doubt the stationarity of most road profiles; however, the assumption concerning the uniformity of traversal velocity is evidently not strictly true and it is usually introduced for simplification of the analysis. Often, the actual velocity of travelling systems is variable. Sobczyk and Macvean [105] have shown that a variable traversal velocity generates nonstationarity in the excitation of the vehicle system, so the responses (displacements, stresses, etc.) are also nonstationary random processes. Another reason for considering nonstationary loadings (applied stresses) is a need to account for the transient dynamic behavior of randomly vibrating structures (subjected to stationary loadings).

As we mentioned in Section 13.2, the problem characterizing the crossing of a given *level* that is represented by a nonconstant curve $u(t)$ can be regarded as a zero-crossing problem for a nonstationary process. Indeed, if $S(t)$ is a stationary process, the number of times $S(t)$ crosses the curve $u(t)$ in time interval $[0, T]$ is precisely the number of times that the process $\underset{\sim}{S}(t) = S(t) - u(t)$ crosses the zero level at that time. Clearly, the crossing characteristics of nonconstant *levels* are of great importance in fatigue reliability (e.g., critical fatigue crack length depends on the loading process and material deterioration during fatigue accumulation, so it should be regarded as a function of time).

Let $S(t)$ be a nonstationary Gaussian random process with mean $m_S(t)$, covariance function $K_S(t_1, t_2)$, and continuous sample functions. Let $N_0(0, T)$ denote the number of crossings of an arbitrary continuously differentiable curve $u(t)$ in the interval $[0, T]$. The average value of $N_0(0, T)$ is given by (see [61])

$$\langle N_0(0, T)\rangle = \int_0^T \sigma_{\dot{S}}(t) \frac{1}{\sigma_S(t)} (1 - \mu^2(t))^{1/2} \varphi\left(\frac{m_S(t)}{\sigma_S(t)}\right) \times [2\varphi(\eta(t)) + \eta(t)(2\Phi(\eta(t)) - 1)]\, dt, \tag{3.42}$$

where

$$\sigma_{\dot{S}}^2(t) = \left.\frac{\partial^2 K_S(t_1, t_2)}{\partial t_1 \partial t_2}\right|_{t_1 = t_2 = t}, \tag{3.43}$$

$$\mu(t) = \frac{\operatorname{cov}[S(t), \dot{S}(t)]}{\sigma_S(t)\,\sigma_{\dot{S}}^2(t)} = \frac{1}{\sigma_S(t)\,\sigma_{\dot{S}}^2(t)} \left.\frac{\partial K_S(t_1, t_2)}{\partial t_2}\right|_{t_1 = t_2 = t}, \tag{3.44}$$

$$\eta(t) = \frac{\dot{m}_S(t) - \dfrac{\sigma_{\dot{S}}^2(t)\,\mu(t)\,m_S(t)}{\sigma_S(t)}}{\sigma_{\dot{S}}(t)\,[1 - \mu^2(t)]^{1/2}}. \tag{3.45}$$

The formula for the mean number of crossings of the curve $u(t)$ is directly obtainable from Eq. (3.42). It is only necessary to replace $m_S(t)$ with $m_S(t) - u(t)$ and $\dot{m}(t)$ with $\dot{m}(t) - \dot{u}(t)$; as we know, adding or subtracting a deterministic function does not affect the covariance function, so $\sigma_{\dot{S}}^2(t)$, $\mu(t)$ remain the same.

If $S(t)$ is a stationary Gaussian process with zero mean and finite second spectral moment and if $u(t)$ is continuously differentiable, then $\sigma_{\dot{S}}^2 = \lambda_2$, $\mu(t) = 0$ (since $S(t)$ and $\dot{S}(t)$ are uncorrelated for each t), and $\eta(t) = -\dot{u}(t)/\sqrt{\lambda_2}$. Therefore, the following equation for the mean number of crossings of the curve $u(t)$ by a stationary Gaussian process is obtained,

$$\langle N_{u(t)}(0, T)\rangle = \frac{1}{\sigma_S}\int_0^T \varphi\left(\frac{u(t)}{\sigma_S}\right)\left\{2\sigma_{\dot{S}}\varphi\left(\frac{\dot{u}(t)}{\sigma_{\dot{S}}}\right) + \dot{u}(t)\left[2\Phi\left(\frac{\dot{u}(t)}{\sigma_{\dot{S}}}\right) - 1\right]\right\} dt. \tag{3.46}$$

For a straight line barrier, $u(t) = a + bt$, $b \neq 0$, the preceding equation reduces to

$$\langle N_{u(t)}(0, T)\rangle = \left(\frac{\sigma_{\dot{S}}}{b}\right)\left[2\varphi\left(\frac{b}{\sigma_{\dot{S}}}\right) + \frac{b}{\sigma_{\dot{S}}}\left(2\Phi\left(\frac{b}{\sigma_{\dot{S}}}\right) - 1\right)\right]\left[\Phi\left(\frac{a+b}{\sigma_S}\right) - \Phi\left(\frac{a}{\sigma_S}\right)\right]. \tag{3.47}$$

As far as maxima of a nonstationary process are concerned, the average number of maxima above the level u for a general nonstationary process is given by Eq. (3.37). This equation can be evaluated when the joint probability density of $S(t)$, $\dot{S}(t)$, and $\ddot{S}(t)$ at time t is given. If the process $S(t)$ is Gaussian, the evaluation is straightforward.

Closing this section, we wish to mention that serious efforts have been made to extend the concept of an envelope to nonstationary processes, (See Yang [110], Krenk [88], and Krenk et al. [89].)

16. Cycle Counting

The results presented in previous sections — particularly these associated with the average number of crossings or peaks — take particularly simple forms in the case of a stationary narrow-band process; in this case, each up-crossing of the mean level corresponds to exactly one local maximum, and a stress cycle is defined as a part of the process between two consecutive up-crossings of the mean level. Stress ranges vary from cycle to cycle and follow a Rayleigh distribution.

As the load bandwidth increases, numerous small peaks occur that obscure the more slowly varying load cycles and make the frequency content of the process considerably more complicated. In this (wide-band) case, there is no obvious definition of stress cycles, and the stress range distribution cannot be directly determined.

To predict the life of a component subjected to complex load histories — including wide-band random loadings — a loading is usually reduced to a sequence of events that can be regarded as compatible with constant-amplitude fatigue data. The methods that make such reductions possible are known as *cycle counting* techniques. (See Dowling [85], where a summary of cycle counting methods is provided.)

Three different methods of cycle counting are most common; namely, *peak counting*, *range counting*, and *rain-flow counting*. The method that has received the widest acceptance in the analysis of fatigue under irregular loading is the rain-flow method. (See [3] and [14].) This method uses a specific cycle counting scheme to account for effective stress ranges and identifies stress cycles related to closed hysteresis loops in the stress–strain response of a material subjected to cyclic loading. It determines the number n and magnitudes $H_i\,(i=1,2,\ldots,n)$ of stress ranges from a stress history. At present, a number of computer algorithms for rain-flow cycle counting have been developed. (See Downing and Socie [86].) Prediction of high cycle fatigue under random loading with the use of the rain-flow method and Monte Carlo simulation has been given by Wirsching and Shehata [109].

It should be emphasized that, though the rain-flow cycle counting method is now regarded as one of the most effective tools for predicting fatigue life under complicated stress histories, it is not able to account for stress interaction effects (i.e., sequence effects). Moreover, because of the heuristic nature of standard rain-flow counting techniques, as well as their complicated *sequential* structure, it is difficult to determine the probability distribution of the rain-flow amplitudes for a random loading process.

In recent years, serious attempts have been made to formulate the rain-flow method in a rigorous mathematical framework. A new definition of the rain-flow cycle amplitude has been given by Rychlik [102] that expresses the rain-flow cycle amplitudes in an explicit analytical manner and provides the basis for deriving the long-time distribution of the rain-flow cycle amplitude for ergodic stationary, Gaussian load processes. (See Rychlik [103], and Lindgren and Rychlik [93].) Rychlik's definition of the rain-flow cycle amplitude is as follows.

Let $s(\tau)$, $-T < \tau < T$, be a single trajectory of the load, and suppose that it has a local maximum at time t (with height $u = s(t)$). Let t^+ be the time of the first up-crossing after t of the level $u = s(t)$, or $t^+ = T$ if no such up-crossing exists in $\tau \in (t, T)$, and let t^- be the time of the last down-crossing before t of the level $u = s(t)$, or $t^- = -T$ if no such down-crossing exists in $\tau \in (-T, t)$. Let us define

$$
\begin{aligned}
H^-(t) &= \max_{t^- < \tau < t} \{s(t) - s(\tau)\}, \\
H^+(t) &= \max_{t < \tau < t^+} \{s(t) - s(\tau)\}.
\end{aligned}
\tag{3.48}
$$

The rain-flow cycle amplitude $H(t)$ at time t is defined as (Fig. 3.4)

$$
H(t) = \min[H^-(t), H^+(t)]. \tag{3.49}
$$

Of course, the times t^+ and t^- depend on u. The equivalence of the preceding definition and the standard rain-flow cycle amplitude has been demonstrated by Rychlik in [102].

It should be observed that Eqs. (3.48) and (3.49) imply that the rain-flow cycle amplitude $H(t)$ is greater than some fixed value h if and only if $s(t)$ reaches the level $u - h$ on both sides of t before it exceeds u again. This indicates that the distribution of H is related to the solution of a first passage time problem in the sequence of maxima and minima.

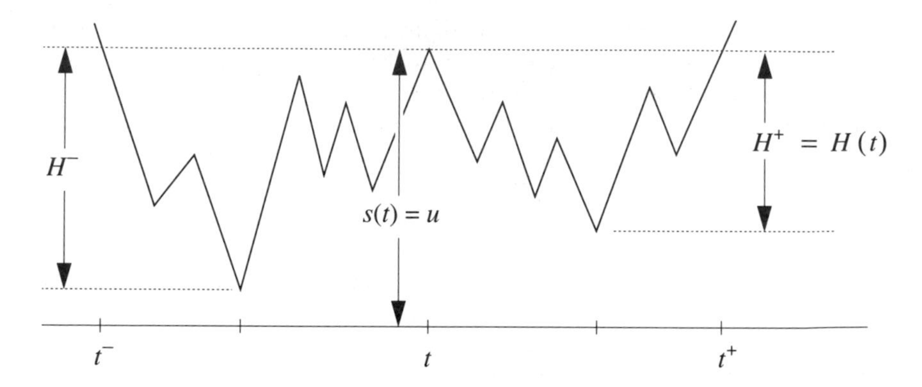

Figure 3.4. Illustration of the rain-flow cycle amplitude [104].

Let $\{t_k\}$ be a sequence of times of the local maxima of $s(t)$ such that $\ldots < t_{-1} < 0 < t_0 < t_1 < \ldots$. The empirical joint probability distribution of $H(t)$ and the local maximum value M_0 at t can be defined as

$$F_{H,M_0}(h, x;T) = \frac{\text{number}\{t_i \in (-T,T); H(t_i) \leq h, s(t_i) \leq x\}}{\text{number}\{t_i \in (-T,T)\}}. \quad (3.50)$$

To obtain the theoretical distribution of (H, M_0), one should take the limit of Eq. (3.50) as $T \to \infty$. It turns out that for some functions $s(t)$, the distribution $F_{H,M_0}(h, x;T)$ can be divergent. However, when $s(t)$ is a trajectory of an ergodic process, the limit of $F_{H,M_0}(h, x;T)$ as $T \to \infty$ exists and defines the joint distribution function $F_{H,M_0}(h, x)$.

The conditional distribution of the rain-flow cycle amplitude $H(t)$ given $M_0 = u$ can be defined as

$$F_{H|M_0}(h|u)$$
$$= 1 - P\{M_k \text{ crosses } (u-h) \text{ before crossing } u, k \to \pm\infty | M_0 = u\}, \quad (3.51)$$

where M_k denotes the local extremum (maximum or minimum) after t, if $k > 0$, or before t if $k < 0$. Unfortunately, the exact evaluation of the aforementioned distribution requires knowledge of all multi-dimensional distributions of the sequence $\{M_k\}$, which usually are not available. Hence, appropriate approximations must be introduced. The approximation that leads to effective results is based on the assumption that the sequence of extrema has some type of Markov structure.

The general case occurs when the extreme values $\{M_k\}$ of the loading process are assumed to form an n-step Markov chain. (Dependence is extended to those M_k, M_j for which $|k-j| \leq n$.) In this case, the distribution of the rain-flow cycle amplitude can be obtained as a solution of a linear integral equation of Fredholm type. (See [103].) If the sequence of extrema is assumed to form a one-step Markov chain with a finite number of states, then the integral equation simplifies to a linear system of algebraic equations. In this case, the transition probabilities of $\{M_k\}$ are determined from the joint probability density of $[M_0, M_1]$; this, however, requires some approximations.

Two questions are of prime interest. The first one can be stated as follows: What is the class of real stochastic processes whose extrema constitute a Markov chain? The second question is: How one can characterize a Markov chain (transition probabilities) of extrema for standard stochastic processes, such as stationary and Gaussian ones? The first question seems to be an open problem. The second question is discussed by Rychlik in [103], whereas Krenk and Gluver [90] construct a one-step Markov matrix for fast, direct simulation of the extremes of a Gaussian loading process.

Chapter IV

Random Fatigue: Evolutionary Probabilistic Models

17. Introductory Remarks

A general probabilistic approach to modelling the fatigue process can be obtained if we consider a sample of material and the fatigue accumulation taking place in it as a dynamical system whose evolution in time is represented by a stochastic vector process $\mathbf{X}(t) = [X_1(t), \dots, X_n(t)]$. The component processes $X_i(t)$, $i = 1, \dots, n$ characterize specific features of fatigue accumulation, where $0 \le X_i(t) \le x_i^*$, and x_i^* denotes the critical or limiting value of $X_i(t)$.

To obtain a model for the random fatigue process, one should deduce the time-varying behavior of the process $\mathbf{X}(t)$ from knowledge of the fatigue process in actual materials. Two possible approaches can be adopted. (See Sobczyk [128].) The first approach seeks to directly describe the evolution of the probability distribution of $\mathbf{X}(t)$ in time and will be discussed in the present chapter. Alternatively, one could construct and analyze the process $\mathbf{X}(t)$ by viewing it as a random function of time. This second approach will be discussed in subsequent chapters.

Denoting the probability density of the transition of the process $\mathbf{X}(t)$ from the state $\mathbf{x}_0$ at t_0 to the state $\mathbf{x}$ at time t by $p(\mathbf{x}, t) = p(\mathbf{x}, t | \mathbf{x}_0, t_0)$, we can postulate an evolutionary-type equation of the form

$$\frac{\partial p(\mathbf{x}, t)}{\partial t} = \mathcal{L}[p(\mathbf{x}, t); \mathbf{s}, \mathbf{z}], \tag{4.1}$$

where $\mathcal{L}$ is termed the fatigue operator (it can be a differential, integral, or differential–integral operator), **s** denotes a vector of appropriate characteristics of the external loading, and **z** denotes all other possible quantities affecting the fatigue process. In such a formulation, the modelling problem lies in proper construction of the fatigue operator $\mathcal{L}$. This, however, is not an easy problem.

To utilize the preceding ideas, one has to introduce several assumptions and hypotheses regarding the process $X(t)$. A hypothesis that has been often introduced requires the fatigue process to be *Markovian*. This makes it possible to use a large variety of mathematical schemes elaborated in the theory of Markov stochastic processes. Since the Markovian property requires that the *future* of the process depends only on its *present* and is independent of the *past*, one might fear that all history-dependent effects have to be excluded. This problem can be alleviated if we formulate the problem (as was indicated before) within the framework of multi-dimensional Markov processes; the history-dependent effects can then be accounted for by appropriate selection of the state space of the process. Moreover, we should keep in mind that the knowledge of the *present* always contains some *accumulated* information about the past.

In the following sections, we shall discuss the evolutionary modelling of fatigue making use of Markov chains and diffusion Markov processes.

18. Fatigue Accumulation as Markov Chain

18.1 General Concept and Special Cases

Let us consider a sample of material from the point of view of damage accumulation and assume that it can be in one of $n+1$ states $E_0, E_1, \ldots, E_n = E^*$, where E_0 denotes the *initial* state and E^* the state characterizing ultimate damage or *failure*. Only forward transitions from one state to another are possible; that is,

$$E_0 \rightarrow E_1 \rightarrow E_2 \rightarrow \ldots \rightarrow E_k \rightarrow E_{k+1} \rightarrow \ldots E_n = E^* . \qquad (4.2)$$

Let us assume the following probabilistic mechanism of the transition: If at time t the sample is in the fatigue state E_k, the probability of the transition $E_k \rightarrow E_{k+1}$ in the time interval $(t, t+\Delta t)$ is equal to $q_k \Delta t + o(\Delta t)$. The probability of the transition from E_k to a state different from E_{k+1} is $o(\Delta t)$.

Let $P_k(t)$ be the probability that a specimen at time t is in the state E_k. It is easily seen that

$$P_k(t+\Delta t) = P_k(t)\,(1-q_k\Delta t) + P_{k-1}(t)\,q_{k-1}\Delta t + o(\Delta t)\ . \qquad (4.3)$$

By transposing the term $P_k(t)$, dividing by Δt, and taking the limit, we obtain the system of differential equations

$$\frac{dP_0(t)}{dt} = -q_0 P_0(t)\,, \qquad (4.4)$$

$$\frac{dP_k(t)}{dt} = -q_k P_k(t) + q_{k-1} P_{k-1}(t)\,, \quad k \geq 1\,. \qquad (4.5)$$

The initial condition says that at time $t = 0$, the system was in the state E_0, that is

$$P_0(0) = 1\,. \qquad (4.6)$$

Equations (4.4) and (4.5), together with Eq. (4.6), completely describe the considered process.

If fatigue is characterized by the length of a dominant crack in a structural element, then the states E_k, $k = 0, 1, \ldots, n$, simply denote the crack tip position at time t measured by integer numbers. Since $E_n = E^*$ is the final state of the sample (i.e., ultimate damage), further transitions are impossible; this indicates that $q_n = 0$.

Examining Eqs. (4.4) and (4.5), one can easily see that the probability of a change in the interval $(t, t+\Delta t)$ depends only on the state in which the system is at time t. Therefore, the prescribed process is Markovian and is known in the theory of population growth as the pure birth process.

The solution of Eq. (4.4) is given by

$$P_o(t) = e^{-q_0 t} > 0\,. \qquad (4.7)$$

For a given q_k, one easily obtains the recursion relation

$$P_k(t) = q_{k-1} e^{-q_k t} \int_0^t e^{-q_k x} P_{k-1}(x)\,dx\,, \quad k = 1, 2, \ldots. \qquad (4.8)$$

The probability of failure at time t is given by the preceding formula when $k = n$.

To obtain more concrete results, one has to make several assumptions concerning the intensities q_k. If the states E_k are characterized by the integers (e.g., this can be assumed when damage is measured by the length of a dominant crack), then the following linear dependence seems to be acceptable (linear birth process):

$$q_k = k\lambda^0, \quad k \geq 1, \lambda^0 > 0. \tag{4.9}$$

This implies that the probability of a transition in the interval $(t, t + \Delta t)$ is proportional to the fatigue damage state (i.e., amount of damage), with the constant of proportionality being λ^0. In this case, Eq. (4.5) becomes

$$\frac{dP_k(t)}{dt} = -k\lambda^0 P_k(t) + \lambda^0 (k-1) P_{k-1}(t), \quad k \geq 1, \tag{4.10}$$

and one obtains the Furry–Yule distribution given by

$$P_k(t) = e^{-\lambda^0 t}(1 - e^{-\lambda^0 t})^{k-1}, \quad k \geq 1, \tag{4.11}$$

which is a monotonically decreasing function of the damage index k.

If the transition probability is the same for all states, i.e., $q_k = \lambda_0$, then from Eqs. (4.7) and (4.8), one obtains the Poisson distribution by induction,

$$P_k(t) = \frac{(\lambda_0 t)^k}{k!} e^{-\lambda_0 t}. \tag{4.12}$$

When $k = n$, Eq. (4.12) yields the probability of failure at time t, which is also the probability that the system changed its fatigue state n times during a time interval of length t. Defining t_k as the time at which the fatigue process enters state k, then the random variables $\xi_1 = t_1 - t_0$, $\xi_2 = t_2 - t_1$, $\ldots, \xi_n = t_n - t_{n-1}$ characterize the intervals between events in the Poissonian stream of fatigue damage states. The random variable $S_n = \xi_1 + \xi_2 + \ldots + \xi_n$ is equal to the *waiting time* $t_n - t_0$ for event $E_n = E^*$. The event $\{S_n \leq t\}$ indicates that at time t, at least n fatigue transitions have taken place. Therefore, Eq. (4.12) with $k = n$ is the distribution function of S_n. Since $\Sigma P_k(t) = 1$, we have

$$P(S_n \leq t) = 1 - \sum_{k=0}^{n-1} \frac{(\lambda_0 t)^k}{k!} e^{-\lambda_0 t}. \tag{4.13}$$

Taking the derivative of this equation with respect to t gives the probability density function of the lifetime of the considered system, i.e.,

$$f(t) = \lambda_0 \frac{(\lambda_0 t)^{n-1}}{(n-1)!} e^{-\lambda_0 t}, \quad t \geq 0. \tag{4.14}$$

The distribution given by Eq. (4.14) belongs to the family of gamma distributions. This result is in agreement with many experimental observations.

The reasoning just presented is based on the work of Sobczyk [127]. An analogous formulation was presented by Ghonem and Provan [119] in which micromechanical interpretation of the fatigue accumulation process was attempted. Modelling of fatigue by Markov chains can also be found in the work of Nemec and Sedlacek [124].

The only quantity needed for calculation of failure probabilities is the transition intensity q_k, which should be estimated from experimental data. Of course, the fatigue experiments should be adequately designed so that appropriate data can be obtained for model parameter estimation, and methods of statistical inference for pure birth process and Poisson process can be adopted. (See Keiding [121], and Misra and Sorenson [122]). Though the model process discussed before is simple, the transition intensities q_k are, in general, different for different *fatigue states*; therefore, Eq. (4.8) can be a useful relation when it is applicable.

18.2 Duty-Cycle Markov Chain Model

A detailed elaboration of the idea of modelling of fatigue accumulation by use of Markov chain theory has been given in a series of papers by Bogdanoff, Krieger and Kozin [111], [112], [113], [114]. The model presented is a discrete-time and discrete-state Markov chain, and the basic notion is a so- called *duty-cycle*. The most essential features of this model are as follows.

Let the discrete damage states be denoted by $x = 1, 2, ..., n$, where state n denotes failure. A basic concept in the model is a duty-cycle (DC), which is understood to be a repetitive period of operation in the life of a component during which damage can accumulate. For constant-amplitude loading, a duty cycle can correspond to a certain number of load cycles; in the case of aircraft operation, each mission can be divided into duty cycles for taxiing, takeoff, landing, etc. Time (discrete) $t = 0, 1, 2, \ldots$ is measured in number of DCs. How damage is accumulated within a DC is not a matter of concern. The increment in damage (which takes place only at the end of each DC) is assumed to depend in a probabilistic manner only on the DC itself and the value of the damage accumulated at the start of that duty cycle. The increment in damage is, however, independent of how the damage was accumulated up to the start of the duty cycle. These assumptions are Markovian ones, and the damage process is regarded as a discrete-space and discrete-time Markov chain.

A Markov chain can be defined by specifying the initial distribution $\boldsymbol{P}_0$ and the transition matrix $\underset{\sim}{\mathbf{P}} = \{P_{ij}\}$. The initial state of damage is specified by the vector $\{\pi_i\}$, where π_i is the probability of damage in state i at $t = 0$, i.e.,

$$\mathbf{P}_0 = [\pi_1, \pi_2, \ldots, \pi_n], \quad p\pi_i \geq 0, \quad \sum_{i=1}^{n} \pi_i = 1. \tag{4.15}$$

The transition matrix for a duty cycle is $\{P_{ij}\}$, where P_{ij} is the probability that the damage is in state j after the duty cycle, provided the damage was in state i at the beginning of the duty cycle. The state of damage at time t is given by the vector

$$\mathbf{P}_t = [P_t(1), P_t(2), \ldots, P_t(n)], \quad P_t(i) \geq 0, \quad \sum_{i=1}^{n} P_t(i) = 1, \tag{4.16}$$

where $P_t(i)$ is the probability that damage is in state i at time t. Markov chain theory gives the relation

$$\mathbf{P}_t = \mathbf{P}_0 \underset{\sim}{\mathbf{P}}_1 \underset{\sim}{\mathbf{P}}_2 \ldots \underset{\sim}{\mathbf{P}}_{t-1} \underset{\sim}{\mathbf{P}}_t, \tag{4.17}$$

where $\underset{\sim}{\mathbf{P}}_j$ is the transition matrix for the jth duty cycle. The expression (4.17) completely specifies the probability distribution of damage at any time. If the damage cycles are of the same severity, Eq. (4.17) reduces to

$$\mathbf{P}_t = \mathbf{P}_0 \underset{\sim}{\mathbf{P}}^t, \tag{4.18}$$

where $\underset{\sim}{\mathbf{P}}$ is the common transition matrix. Since matrix multiplication is generally not commutative, it follows from Eq. (4.17) that the order of the duty cycles is important for damage accumulation within this model.

The probability distributions of various random variables associated with the damage accumulation process can be easily determined. For example, the time to failure T_n (i.e., to reach absorbing state n) has the distribution function $F_n(t)$ given by

$$F_n(t) = P\{T_n \leq t\} = P_t(n), \quad t = 1, 2, \ldots . \tag{4.19}$$

Since the model just characterized has been widely described in the literature, including a book by Bogdanoff and Kozin [115] and the book by Madsen et al. [31], the interested reader is referred to these works and to the original papers for additional details [111], [112], [113], [114]. The authors show that this model has many advantages. However, since the physics of the fatigue phenomenon are not sufficiently taken into account (i.e., there is no direct relation between damage and measurable physical quantities), model identification, or parameter estimation, is rather difficult. (See Madsen et al. [31].)

18.3 Modelling of Short Crack Growth

An interesting approach to modelling of fatigue by use of Markov chains has been reported by Cox and Morris [116], [117]. The authors make an attempt to draw a link between the microstructure of material and the statistics of fatigue growth rates of short cracks. The basic idea is as follows.

In addition to the crack length A, the so-called *growth control variable* u is introduced; this variable is regarded as a measure of some local physical property of the material that varies randomly and controls the crack growth at each point. The growth control variable is open to a variety of interpretations, depending on the mechanisms known to control crack growth in specific situations.

It is assumed that the evolution of u is described by a finite-state Markov chain indexed on discrete crack length, and that the evolution of the crack length A is governed by a first-order differential equation. The Markov chain characterizing the control variable u is defined by the probability transition matrices $\underset{\sim}{\mathbf{P}}_m$, $m = 0, 1, \ldots$. The subscript m refers to the mth value A_m of the discretized crack length. The element $(P_m)_{ij}$ of $\underset{\sim}{\mathbf{P}}_m$ denotes the probability that the discretized variable u is in the jth state when the crack length is A_m, given that it was in the ith state when the crack length was A_{m-1}. The assumed Markovian property implies that $\underset{\sim}{\mathbf{P}}_m$ does not depend on the history of the crack prior to it reaching length A_{m-1}. The calculation of the elements $(P_m)_{ij}$ requires a specific model for crack growth; it may involve empirical parameters to be established by comparing the output of the model with crack growth data.

The probabilities $P_m(i)$ that the growth control variable u has the value u_i at crack length A_m form the vector,

$$\mathbf{P}_m = [P_m(1), P_m(2), \ldots, P_m(M)], \tag{4.20}$$

where M is the number of discrete values of u. (It is assumed that M does not depend on m.) If $\mathbf{P}_0$ denotes the initial distribution of u, then

$$\mathbf{P}_m = \mathbf{P}_0 \prod_{i \le m} \underset{\sim}{\mathbf{P}}_i. \tag{4.21}$$

The preceding formula enables one to calculate the mean and moments of u at any crack length. The mean $\langle u_m \rangle$ and the standard deviation σ_{u_m} of u at crack length A_m are given by

$$\langle u_m \rangle = \sum_{j=1}^{M} u_m(j) P_m(j) , \tag{4.22}$$

$$\sigma_{u_m} = \left[\sum_{j=1}^{M} u_m^2(j) P_m(j) - \langle u_m \rangle^2 \right]^{1/2} . \tag{4.23}$$

A cumulative probability distribution of u at crack length A_m is given by the formula,

$$F_u(u_m(j)) = \sum_{i \le j} P_m(i) . \tag{4.24}$$

The preceding quantity can be used to generate the distribution of crack growth rates at the length A_m. Equation (4.20) can also be used to calculate the covariance between the values of u at two different crack lengths, which should be a decreasing function of the separation of two lengths of a crack $|A_m - A_k|$.

To make the model effective, the rate of crack growth should be specified as a function of the applied stress and the growth control variable u. Assume that such an equation is known or postulated and that it has the form,

$$\frac{dA}{dt} = f(A, u) . \tag{4.25}$$

The average value of the crack growth rate at a crack length A_m is given by

$$\langle f_m \rangle = \sum_{j=1}^{M} f(A_m, u_m(j)) P_m(j) . \tag{4.26}$$

The correlation between the crack growth rates at two different crack lengths is characterized by the quantity,

$$\langle f_m f_k \rangle = \sum_{i,j=1}^{M} f(A_m, u_m(i)) f(A_k, u_k(j)) P_m(i) P_k(j) . \tag{4.27}$$

The preceding quantity will decay with increasing $|A_m - A_k|$. If this decay is characterized by an exponential function, a characteristic correlation length can be introduced that reflects some intrinsic length of scale of the material. The model presented by Cox and Morris [116], [117], described before, provides the possibility to relate this length scale with statistics of crack growth.

In [116], the authors present an analysis of statistical data taken for short surface fatigue crack (of length 10–250 μm) based on the model characterized before. Various *models* for the growth control variable u are discussed; they are consistent with [120], in which the so-called roughness-induced closure is found to be the dominant controlling factor of fluctuations in the growth rate. Also, use is made of an empirical crack growth relation of the form of Eq. (4.25). Further analysis of the model, including comparisons between predicted probability distributions and their empirical counterparts, would undoubtedly increase the value of this new idea.

19. Markov Diffusion Models

19.1 General Approach

Although fatigue accumulation (e.g., fatigue crack growth) evolves in time intermittently, and in reality constitutes a discontinuous random process, it is reasonable to approximate it by a continuous stochastic process. Such modelling of discontinuous phenomena has brought about interesting results in other fields (e.g., in population biology, physics, and chemistry). For example, in the study of the motion of particles in a colloidal suspension, this approach leads to the diffusion equation that governs the transition density of typical particles in a colloidal suspension. It is, therefore, of interest to regard the fatigue crack growth process as a diffusion-like process. Such an approach to modelling of fatigue has been suggested by Oh [125]; it can be interpreted as another illustration of evolutionary probabilistic modelling.

Let us denote the crack tip position (or crack length) at time t by $A(t)$ and assume that it is a continuous random process. Because crack growth is irreversible, $A(t)$ may only take its values in the interval $[a_0, \infty)$, where $a_0 = A(t_0)$. To put further analysis into the frame of diffusion Markov process theory, we need to assume that growth of the crack is dependent only upon its current crack length. This Markovian assumption must be regarded as a hypothesis, calling for experimental verification. (In Chapter VI, a fracture mechanics-based theoretical justification of this hypothesis will be indicated for a certain class of problems.) Heuristically, we can attempt to motivate the Markovian hypothesis taking into account the basic factors causing crack growth. In the case of deterministic cyclic loading, there are two main factors: the range of stress intensity factor and the strength distribution of the material in the crack tip region. The first of these two factors is a deterministic quantity when the tip position is given, whereas the second is usually not altered by the past positions of the crack tip.

Let $p(a, t|a_0, t_0)$ be the density of the transition probability of the crack tip position. Assuming further that $A(t)$ constitutes a Markov diffusion process, and adopting reasoning common in the theory of such processes, we come to the following governing equation for $p(a, t) = p(a, t|a_0, t_0)$:

$$\frac{\partial p}{\partial t} = -\frac{\partial}{\partial a}[m(a)p(a, t)] + \frac{1}{2}\frac{\partial^2}{\partial a^2}[b(a)p(a, t)], \tag{4.28}$$

where $m(a)$ and $b(a)$ are the coefficients characterizing infinitesimal properties of the process, i.e., the drift and diffusion coefficients, respectively. To be able to use the model, the coefficients $m(a)$ and $b(a)$ have to be properly determined. Using a weakest-link model for the prediction of fatigue crack growth rate, Oh [126] found the following forms of the coefficients:

$$m(a) = C_m a^{1+1/m}, \qquad b(a) = C_b a^{2+1/m}, \tag{4.29}$$

where C_m and C_b are constants given by

$$C_m = \frac{\int_0^1 x\left[\ln\left(1+\frac{c}{c+x}\right)\right]^m dx}{\sigma_0\left[\int_0^1\left[\ln\left(1+\frac{c}{c+x}\right)\right]^m dx\right]^{1-1/m}}\Gamma\left(1-\frac{1}{m}\right)\left[\frac{1}{32}\pi\left(\frac{\sigma_\infty}{\sigma_y}\right)^2\right]^{1+\frac{1}{m}}, \tag{4.30}$$

$$C_b = \frac{\int_0^1 x^2\left[\ln\left(1+\frac{c}{c+x}\right)\right]^m dx}{\sigma_0\left[\int_0^1\left[\ln\left(1+\frac{c}{c+x}\right)\right]^m dx\right]^{1-1/m}}\Gamma\left(1-\frac{1}{m}\right)\left[\frac{1}{32}\pi\left(\frac{\sigma_\infty}{\sigma_y}\right)^2\right]^{2+1/m}, \tag{4.31}$$

and $c = \pi(1-\nu^2)/32$. The integrals in Eqs. (4.30) and (4.31) must be computed numerically.

Within such a model, the probabilistic properties of the diffusion crack growth process can be obtained as a solution of the partial differential equation (4.28) known as the Fokker–Planck–Kolmogorov equation. This equation should be supplemented by the appropriate initial and boundary conditions.

If the initial crack length is regarded to be a deterministic quantity equal to a_0, then

$$p(a, t_0) = p(a, t_0 | a_0, t_0) = \delta(a - a_0), \tag{4.32}$$

where $\delta(\cdot)$ is the Dirac delta. If $A(t_0)$ is a random variable with probability density $p_0(a)$, then the initial condition is

$$p(a, t_0) = p_0(a). \tag{4.33}$$

Since the fatigue crack growth process is considered on the finite interval $[a_0, a^*] \subset [a_0, \infty)$, where a^* denotes a critical crack size, the *boundary conditions* for Eq. (4.28) should be of particular concern.

In general, the boundary conditions of the Fokker–Planck–Kolmogorov equation may be different depending on the behavior of the process at the boundary points. (See Feller's classification of boundaries associated with one-dimensional diffusion process [118].) The point is that, depending on the infinitesimal characteristics of $m(a, t)$ and $b(a, t)$, the process can exhibit various types of behavior at the boundaries of the interval. For example, the coefficients can be such that the process never reaches the boundaries; in this case, no boundary conditions need to be imposed. (These boundaries are called natural.) Another possibility is that coefficients may be such that the process can reach the boundary in a finite amount of time with nonzero probability (accessible boundaries); for such processes, classification of the boundaries and selection of appropriate conditions are essential.

In the context considered here (i.e., the fatigue crack growth process), the conditions that seem to be physically most appropriate are: a reflecting boundary condition at $a = a_0$, and a zero derivative boundary condition at $a = a^*$.

A *reflecting boundary condition* means that flow of the *probability mass* through $a = a_0$ is not allowed; if a trajectory reaches the state $a = a_0$, it is instantaneously *reflected* to the interior of the interval where it is again governed by Eq. (4.28). Analytically, this condition takes the form [60],

$$m(a)p(a, t) - \frac{1}{2}\frac{\partial}{\partial a}[b(a, t)p(a, t)] = 0, \quad \text{at } a = a_0. \tag{4.34}$$

This condition assures that

$$\int_{a_0}^{\infty} p(a, t)\, dt = 1, \tag{4.35}$$

i.e., all probability mass is concentrated on the open interval $[a_0, \infty)$.

The boundary condition at $a = a^*$ is given on physical grounds by Oh [125]. When the crack length a reaches the critical crack size a^*, it is assumed to have an infinite velocity (i.e., unstable crack growth). Therefore, one can argue that

$$p(a^* + \Delta a, t) = p(a^*, t), \tag{4.36}$$

which implies

$$\left.\frac{\partial p}{\partial a}\right|_{a^*} = \lim_{\Delta a \to 0} \frac{p(a^* + \Delta a, t) - p(a^*, t)}{\Delta a} = 0. \tag{4.37}$$

The solution of the Fokker–Planck–Kolmogorov equation (4.28) satisfying the prescribed conditions provides basic information concerning the evolution in time of the process $A(t)$. From the transition density $p(a, t; a_0, t_0)$, one can obtain quantities that are of interest to fatigue life prediction. For example, the probability distribution of the crack size at time t is given by

$$F_A(a, t) = P\{A(t) < a | A(t_0) = a_0\} = \int_{a_0}^{a} p(a, t | a_0, t_0)\, da. \tag{4.38}$$

The probability that the crack size exceeds a given critical level a^* is

$$P\{A(t) > a^* | A(t_0) = a_0\} = 1 - F_A(a^*; t), \tag{4.39}$$

and the probability distribution of random time T_{a^*} when the crack size $A(t)$ reaches a given value a^* is

$$\begin{aligned} F_{T_{a^*}}(\theta) &= P\{T_{a^*} \le \theta\} = 1 - P\{A(t_0 + \theta) < a^*\} \\ &= 1 - \int_{a_0}^{a^*} p(a, t_0 + \theta | a_0, t_0)\, da. \end{aligned} \tag{4.40}$$

We also mention here that an alternative boundary condition may be considered at $a = a^*$. If a^* is such that cracks larger than a^* do not exist (e.g., if a^* is the width of the plate), then an absorbing boundary condition should be prescribed at $a = a^*$. This indicates that trajectories that reach this state are absorbed and will remain there forever. In this situation, the probability $p(a, t)$ should vanish, i.e.,

$$p(a^*, t) = 0. \tag{4.41}$$

Since a diffusion Markov process is characterized by its drift coefficient $m(a)$ and the diffusion coefficient $b(a)$, the proper determination of these functions from empirical data constitutes a crucial issue of modelling. According to the definition of these coefficients, $m(a)$ characterizes the average tendency in the evolution of a process, whereas $b(a)$ reflects an amount of random fluctuations due to unpredictable perturbations of the process. Since fatigue crack growth is a nondecreasing process, the diffusion part in Eq. (4.28) should be sufficiently small. We shall discuss this problem in more detail in Chapter VI in connection with representation of the diffusion fatigue crack growth by stochastic differential equations.

19.2 Numerical Illustration

In this section, we present a numerical example to illustrate the Markov diffusion model [125]. The model parameters are given to be $K_{Ic}/\sigma_y \approx 1.1\ \mathrm{m}^{1/2}$, $\nu = 0.3$, $a_0 = 0.002\ \mathrm{m}$, $\sigma_\infty/\sigma_y = 0.85$, and $a_{cr}/a_0 = 110$. The value of the Weibull parameters in the weakest link model are $m = 1.54$ and $\sigma_0 = 1.411\ \mathrm{m}^{1/2}$. These parameters correspond to the crack growth model

$$\frac{dA}{dN} \approx (\Delta K)^{3.3}, \tag{4.42}$$

where $\Delta K = 16.591 \times 10^6 \mathrm{N/m}^{3/2}$, which is typical for mild steel.

The solution of Eq. (4.28) for these parameters is given in Fig. 4.1 in the form of the evolution of the transition density with time. Note the differences between the scales in the graphs. As shown here, after only four cycles, the probability is concentrated very near the initial position of the crack tip. As time passes, the distribution of crack sizes spreads out and it becomes increasingly likely that the tip will be found away from its initial position. At large times, the probability that the crack has not grown past its initial position has decreased to zero.

Figure 4.2 shows the fatigue life distribution and its corresponding density derived from Eq. (4.40). The slight skewness of the density indicates that it is not Gaussian.

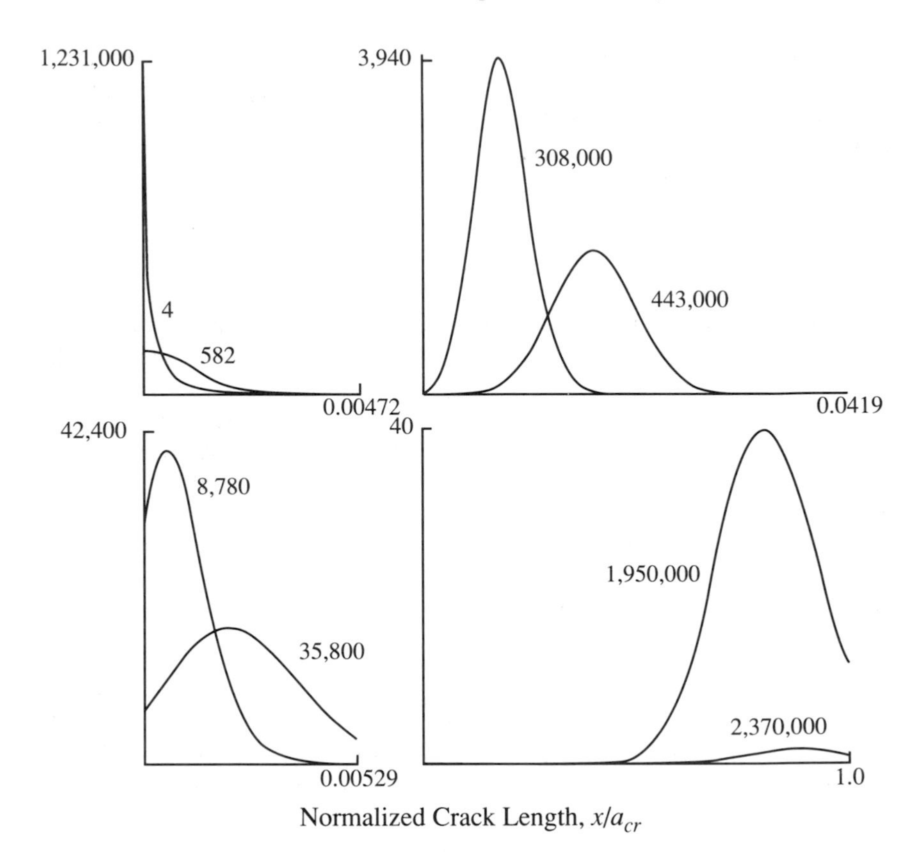

Figure 4.1. Evolution of density function for crack size.

20. Extensions

Modelling of physical phenomena by Markov processes always generates important questions such as: When is the Markovian hypothesis really appropriate? In other words, how can one verify that the future behavior of a physical process does not depend on the past?

As we mentioned at the beginning of this chapter, probabilistic models of physical phenomena may become Markovian, if the dimension of the model (the number of state variables) is made large enough. In such situations, we deal with a multidimensional Markov process $\mathbf{X}(t)$, whose components characterize different features of the phenomenon. If the model process is assumed to be a diffusion Markov process, then the Fokker–Planck–Kolmogorov equation has the form of Eq. (2.119). The development of a suitable state vector and the corresponding evolutionary equa-

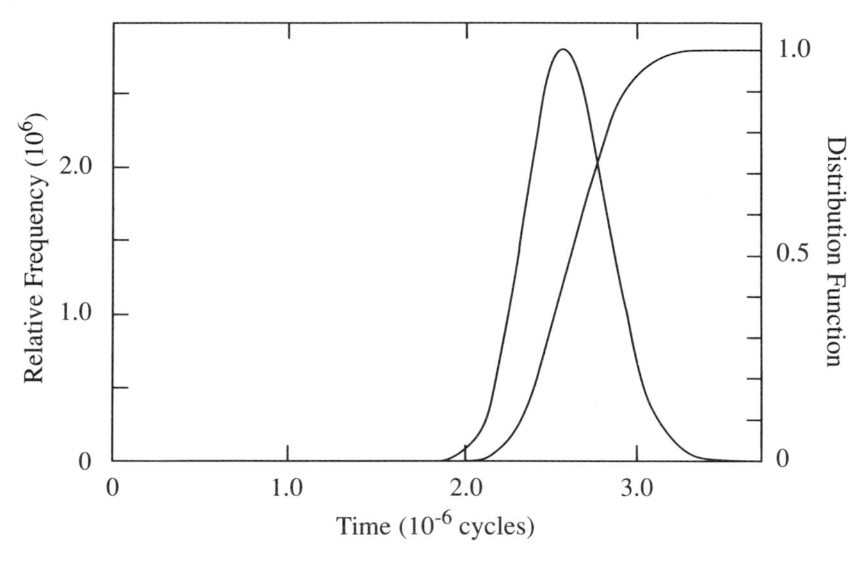

Figure 4.2. Fatigue life distribution.

tions for physical phenomena is a complicated and iterative process. (See [123].) As far as the fatigue process is concerned, at the present time we do not have sufficient data that is suitable for determining the appropriate dimension of a vectorial Markov model or the appropriate state variables. Data on crack length $A(t)$ versus the number of cycles N are usually gathered with load amplitude and other parameters held constant. Either A is recorded as a function of N or N as a function of A. On the other hand, it is known that crack growth is associated with, for example, plastic deformation in the tip of a crack. Therefore, the size of a plastic zone could be considered as a possible second state variable. The temperature (or another thermodynamical quantity) ahead of the crack tip could also be included as an observable. These macroscopic variables, plus others that may be put forward in the future, are possible components to be added to A to constitute a multidimensional Markov process. It seems, as far as we are aware, that at present there are no experimental data obtained from simultaneous observation of the evolution of two or more variables associated with fatigue crack growth. There is no doubt that any progress in this direction would be of great importance.

Chapter V

Random Fatigue Crack Growth: Cumulative Jump Models

21. Introductory Remarks

Many experimental investigations have shown that fatigue cracks grow intermittently and that the growth itself consists of active and dormant periods. During fatigue crack propagation, striations are created that are most often randomly spaced and have considerable statistical dispersion. (See Itagaki and Shinozuka [134].) This is mainly due to the fact that the damaging stresses experienced by the material and causing crack growth are generated by factors such as peaks of the loading process, its rises and falls, etc., which are random and discrete events in time.

In addition to this general mechanism, shocks that occur in separate, discrete instants of time are a common source of time-varying loads. They can, for instance, be generated by explosion-type loads. Loads acting on offshore structures due to sea waves during storm weather conditions also can be regarded as sequences of random impulses.

It is therefore justified to regard the crack growth process as a discontinuous random process consisting of a random number of jumps, each with random magnitude. Although some elements of such reasoning are also present in the Markov chain models, it seems important to treat random elementary crack increments (jumps) in fatigue growth in a more explicit way. The approach to modelling of fatigue crack growth by random sums of random components has recently been elaborated in the papers by Sobczyk and Trebicki [149], [150], and for random crack growth with retardation by Ditlevsen and Sobczyk [133]. In this chapter, we expound on the basic features of the cumulative jump models along with the results relevant to fatigue life prediction.

22. Description of Models

The fatigue process is characterized here by a stochastic process $A(t, \gamma)$, which is regarded as the length of a dominant crack at time t. As before, γ symbolizes the elementary event belonging to the sample space Γ; for each $\gamma \in \Gamma$, $A(t, \gamma)$ represents a possible sample function of the crack growth. The process $A(t, \gamma)$ is represented as a random sum of random components,

$$A(t, \gamma) = A_0 + \sum_{i=1}^{N(t)} Y_i(\gamma), \qquad Y_i(\gamma) = \Delta A_i, \tag{5.1}$$

where A_0 is the initial crack length of sufficient size to propagate (assumed to be known from experimental data); it can be regarded as a deterministic constant or as a random variable. $Y_i(\gamma)$ are the random variables characterizing the magnitudes of elementary crack increments in crack length, and $N(t)$ is an integer-valued stochastic process (a counting stochastic process) characterizing the number of crack increments in the interval $[0, t]$.

The model process $A(t, \gamma)$ constitutes a class of cumulative models in which particular members are specified by the hypotheses posed on the counting process $N(t)$ and on the random variables $Y_i(\gamma)$, $i = 1, 2, \ldots$. Although the counting process $N(t)$ can be taken in a quite general form, to make the model effective we shall assume that $N(t)$ is a Poisson process (homogeneous or inhomogeneous) or a pure birth process. In these cases, the probability $P_k(t)$ that the number of events (crack jumps) is equal to k within the time interval $[0, t]$ is given by Eq. (2.123) for a homogeneous Poisson process, by Eq. (2.126) for an inhomogeneous Poisson process, and by Eq. (2.127) for a pure birth process.

As far as the components $Y_i(\gamma)$ are concerned, we shall assume first (Section 23) that they are independent and identically distributed, nonnegative random variables, and then (Section 24) extension will be presented when $Y_i(\gamma)$ are correlated random variables. Through all analysis, we additionally assume that $N(t)$ is independent of $Y_i(\gamma)$.

In the model given by Eq. (5.1), randomness in the fatigue crack growth process is accounted for via the probabilistic mechanism of the transition from one fatigue state, or crack's length, to another (with Poisson or pure birth intensity) and by the probability distribution of the magnitudes of elementary crack increments.

23. Cumulative Model with Underlying Poisson Process

23.1 Crack Size Distribution

23.1.1 $N(t)$ — Homogeneous Poisson Process

A function of prime interest is the probability distribution of a crack size at arbitrary time t, that is,

$$F_A(a;t) = P\{A(t) \le a\}, \tag{5.2}$$

or equivalently, the probability density function $f_A(a;t)$.

Let

$$A(t, \gamma) = A_0 + A_1(t, \gamma) = A_0 + \sum_{i=1}^{N(t)} Y_i(\gamma). \tag{5.3}$$

To derive an expression for $f_{A_1}(a;t)$, we introduce the moment generating function for $A_1(t, \gamma)$,

$$M(s) = \langle e^{-sA_1(t, \gamma)} \rangle, \tag{5.4}$$

which is the Laplace transform of the probability density function $f_{A_1}(a;t)$ of $A_1(t)$, i.e.,

$$M(s) = \int_0^{\infty} e^{-sa} f_{A_1}(a;t)\, da. \tag{5.5}$$

Assuming that $N(t)$ is a homogeneous Poisson process (with intensity λ_0) and making use of the independence of $Y_i(\gamma)$, we obtain

$$\begin{aligned} \langle e^{-sA_1} \rangle &= \sum_{k=0}^{\infty} \langle e^{-sA_1} | N(t) = k \rangle\, P\{N(t) = k\} \\ &= \sum_{k=0}^{\infty} \langle e^{-sY_i(\gamma)} \rangle^k \frac{(\lambda_0 t)^k}{k!} e^{-\lambda_0 t} \\ &= e^{-\lambda_0 t} \sum_{k=0}^{\infty} [G(s)]^k \frac{(\lambda_0 t)^k}{k!}, \end{aligned} \tag{5.6}$$

where $G(s)$ is the moment-generating function of the random variables $Y_i(\gamma)$.

To make further analysis more efficient, we assume a specific form of the probability distribution of elementary crack increments $Y_i(\gamma)$. The experiments performed by Kogajev and Liebiedinskij and reported in paper [138] indicate that the distribution of elementary crack increments under constant stress intensity factor range can be approximated by the exponential distribution (with the parameter being a random variable). Being supported by these observations, we hypothesize that

$$g(y) = \begin{cases} \alpha e^{-\alpha y}, & y > 0, \alpha > 0 \\ 0, & y \le 0. \end{cases} \tag{5.7}$$

In this case,

$$G(s) = \frac{\alpha}{s+\alpha}, \qquad s > 0. \tag{5.8}$$

The inverse of Eq. (5.6) gives the following result for the density $f_{A_1}(a;t)$,

$$f_{A_1}(a;t) = e^{-\lambda_0 t - \alpha a} \sum_{k=0}^{\infty} \frac{(\alpha\lambda_0 t)^{k+1} a^k}{k!\,(k+1)!}, \qquad a > 0, \tag{5.9}$$

and $f_{A_1}(a;t) = 0$, for $a \le 0$.

The density function in Eq. (5.9) can also be represented in the form,

$$f_{A_1}(a;t) = \sqrt{\frac{\alpha\lambda_0 t}{a}}\, e^{-\lambda_0 t - \alpha a}\, I_1(2\sqrt{\lambda_0 \alpha a t}), \tag{5.10}$$

where $I_1(\cdot)$ is a modified Bessel function of the first order given by

$$I_p(x) = \sum_{k=0}^{\infty} \frac{1}{k!\,\Gamma(k+p+1)} \left(\frac{x}{2}\right)^{2k+p}, \qquad p > -1, \tag{5.11}$$

where $p = 1$. For the properties of $I_p(x)$, see [129]. It is clear that because of Eq. (5.3),

$$f_A(a;t) = f_{A_1}(a - A_0;t). \tag{5.12}$$

Therefore, the mean and variance of $A(t)$ are given by

$$\langle A(t,\gamma)\rangle = \begin{cases} \frac{\lambda_0}{\alpha}t + A_0 , & N(t) > 0 \\ A_0 , & N(t) = 0, \end{cases} \tag{5.13}$$

$$\sigma_A^2 = \frac{2\lambda_0}{\alpha}t , \qquad N(t) > 0. \tag{5.14}$$

23.1.2 $N(t)$ —Inhomogeneous Poisson Process

If the intensity of the transition (of the process $N(t)$) from state k to $k+1$ in the infinitesimal time interval $[t, t+\Delta t]$ is time-dependent (i.e., $\lambda = \lambda(t)$, $t > 0$), then the probability $P_k(t)$ is determined by

$$\begin{aligned} \frac{dP_k(t)}{dt} &= -\lambda(t)P_k(t) + \lambda(t)P_{k-1}(t), \qquad k = 1, 2, \dots , \\ \frac{dP_0(t)}{dt} &= -\lambda(t)P_0(t), \end{aligned} \tag{5.15}$$

with initial conditions

$$\begin{aligned} P_k(0) &= 1 \text{ for } k = 0, \\ P_k(0) &= 0 \text{ for } k = 1, 2, \dots . \end{aligned} \tag{5.16}$$

Introducing the function $\eta(t)$ defined as

$$\lambda(t) = \frac{d\eta(t)}{dt}, \qquad \eta(t) = \int_0^t \lambda(\tau)\,d\tau \tag{5.17}$$

gives the solution

$$P_k(t) = \frac{[\eta(t)]^k}{k!} e^{-\eta(t)} . \tag{5.18}$$

The moment-generating function then takes the form:

$$M(s) = e^{-\eta(t)} \sum_{k=0}^{\infty} \frac{[\eta(t)]^k}{k!} \left(\frac{\alpha}{s+\alpha}\right)^k . \tag{5.19}$$

The Laplace inverse gives for $a > 0$,

$$f_{A_1}(a;t) = e^{-\eta(t) - \alpha a} \sum_{k=0}^{\infty} \frac{[\alpha\eta(t)]^{k+1} a^k}{(k+1)!k!}. \tag{5.20}$$

When $\lambda(t) = \lambda_0$ (i.e., $\eta(t) = \lambda_0 t$), Eq. (5.20) reduces to Eq. (5.9).

Remark. The preceding formulae have been derived under the assumption that A_0 is a deterministic constant. There is no difficulty in obtaining the appropriate distributions when A_0 is a random variable independent of the process $A_1(t, \gamma)$. In such a situation,

$$f_A(a;t) = \int_0^a f_{A_1}(a-x;t) f_{A_0}(x)\, dx, \quad a > 0, \tag{5.21}$$

where $f_{A_1}(a;t)$ is given by Eq. (5.10) or (5.20), and $f_{A_0}(x)$ is the probability density of the initial crack size $A_0(\gamma)$.

23.2 Lifetime Distribution

Let T be a positive random variable that characterizes a random time at which the crack size $A(t, \gamma)$ reaches a fixed critical value ξ. Of course,

$$P\{T > t\} = P\{A(t, \gamma) < \xi\}. \tag{5.22}$$

This means that the lifetime distribution and the distribution of crack size are directly related to each other, namely,

$$F_T(t) = 1 - F_A(a;t)\big|_{a=\xi}. \tag{5.23}$$

The distribution in question in this section can be derived from the results already provided. Nevertheless, it is useful to derive the lifetime distribution independently. According to Eq. (5.22) and assuming that $N(t)$ is a homogeneous process, we obtain

$$\begin{aligned} P\{T > t\} &= P\{A(t, \gamma) < \xi\} = P\{A_0 + \sum_{i=1}^{N(t)} Y_i(\gamma) < \xi\} \\ &= \sum_{k=0}^{\infty} P\{A_0 + \sum_{i=1}^{N(t)} Y_i(\gamma) < \xi | N(t) = k\} P_k(t) \qquad (5.24) \\ &= \sum_{k=0}^{\infty} P_k(\xi) P_k(t) = \sum_{k=0}^{\infty} P_k(\xi) e^{-\lambda_0 t} \frac{(\lambda_0 t)^k}{k!}, \end{aligned}$$

where

$$P_k(\xi) = P\{\sum_{j=1}^{k} Y_i(\gamma) < \xi - A_0\} \tag{5.25}$$

is the probability that after k increments, the crack size is less than critical value ξ. Of course,

$$1 = P_0(\xi) > P_1(\xi) > \ldots > P_k(\xi) > P_{k+1}(\xi) > \ldots . \tag{5.26}$$

Since the distribution function $F_T(t)$ of T is $1 - P\{T > t\}$, we have the following result for the probability density:

$$\begin{aligned} f_T(t) &= \frac{dF_T(t)}{dt} = -\frac{d}{dt}P\{T > t\} \\ &= \lambda_0 e^{-\lambda_0 t} \sum_{k=0}^{\infty} \left[\frac{(\lambda_0 t)^k}{k!} - \frac{k(\lambda_0 t)^{k-1}}{k!} \right] P_k(\xi) \\ &= \lambda_0 e^{-\lambda_0 t} \left[\sum_{k=0}^{\infty} \frac{(\lambda_0 t)^k}{k!} P_k(\xi) - \sum_{k=0}^{\infty} \frac{(\lambda_0 t)^k}{k!} P_{k+1}(\xi) \right] \\ &= \lambda_0 e^{-\lambda_0 t} \sum_{k=0}^{\infty} \frac{(\lambda_0 t)^k}{k!} [P_k(\xi) - P_{k+1}(\xi)] . \end{aligned} \tag{5.27}$$

Let $g(y)$ be a common probability density of the random variables $Y_i(\gamma)$, $i = 1, 2, \ldots, k$; and let

$$Z(\gamma) = \sum_{i=1}^{k-1} Y_i(\gamma) . \tag{5.28}$$

Then

$$\begin{aligned} P_k(\xi) &= P\{Z + Y_k < \xi - A_0\} \\ &= \int_0^{\xi - A_0} P_{k-1}(x)\, g[(\xi - A_0) - x]\, dx, \end{aligned} \tag{5.29}$$

where $P_{k-1}(x)$ is the distribution of $Z(\gamma)$. It was shown [118] that the distribution function of $Y_1 + Y_2 + \ldots + Y_k$ is

$$G_k(y) = 1 - e^{-\alpha y}\left(1 + \frac{\alpha y}{1!} + \ldots + \frac{(\alpha y)^{k-1}}{(k-1)!}\right). \tag{5.30}$$

Since $P_k(\xi)$ in Eq. (5.29) is the distribution of a sum of k independent random variables, then we have

$$\begin{aligned} P_k(\xi) - P_{k+1}(\xi) &= G_k(\xi - A_0) - G_{k+1}(\xi - A_0) \\ &= \frac{a^k (\xi - A_0)^k}{k!} e^{-\alpha(\xi - A_0)}, \end{aligned} \tag{5.31}$$

and the expression (5.27) for the density of random variable T reduces to

$$f_T(t) = \lambda_0 e^{-\lambda_0 t - \alpha(\xi - A_0)} \sum_{k=0}^{\infty} \frac{(\lambda_0 t)^k}{k!} \frac{\alpha^k (\xi - A_0)^k}{k!}, \qquad t > 0. \tag{5.32}$$

It can be easily shown that $f_T(t)$ given by Eq. (5.32) is a valid probability density function. It is nonnegative and, when integrated from zero to infinity, is equal to one. A more useful form of Eq. (5.32) is

$$f_T(t) = \lambda_0 e^{-\lambda_0 t - \alpha(\xi - A_0)} I_0\left(2\sqrt{\lambda_0 t \alpha(\xi - A_0)}\right), \tag{5.33}$$

where $I_0(\cdot)$ is a modified Bessel function of order zero; see Eq. (5.11).

Making use of the definition of the mean value and variance, we obtain after integration

$$\mu_T = \langle T \rangle = \frac{1}{\lambda_0}[\alpha(\xi - A_0) + 1], \tag{5.34}$$

$$\sigma_T^2 = \frac{1}{\lambda_0^2}[2\alpha(\xi - A_0) + 1]. \tag{5.35}$$

It is seen that the mean lifetime $\langle T \rangle$ and the variance of T are proportional to the critical value ξ when the parameters a and λ_0 are fixed. Equation (5.34) also shows that when λ_0 in the Poisson stream of crack jumps decreases, the average lifetime of a specimen increases, which agrees with our expectations.

Remark. If the process $N(t)$ is inhomogeneous, then the probability $P_k(t)$ is given by Eq. (5.18). If we use this expression in Eq. (5.24), the probability density of T is

$$
\begin{aligned}
F_T(t) &= -\frac{d}{dt} P\{T > t\} = -\frac{d}{dt}\left[e^{-\eta(t)} \sum_{k=0}^{\infty} \frac{[\eta(t)]^k}{k!} P_k(\xi) \right] \\
&= \frac{d\eta(t)}{dt} e^{-\eta(t)} \sum_{k=0}^{\infty} \frac{[\eta(t)]^k}{k!} [P_k(\xi) - P_{k+1}(\xi)], \qquad t > 0.
\end{aligned} \tag{5.36}
$$

In the case when $Y_i(\gamma)$ are of exponentially distributed random variables, we have

$$
f_T(t) = \frac{d\eta(t)}{dt} e^{-\eta(t) - \alpha(\xi - A_0)} \sum_{k=0}^{\infty} \frac{\alpha^k (\xi - A_0)^k}{k!} \frac{[\eta(t)]^k}{k!}. \tag{5.37}
$$

It is seen that in the case where $\lambda(t) = \lambda_0$ (i.e., $\eta(t) = \lambda_0 t$), Eq. (5.37) reduces to Eq. (5.32).

23.3 Estimation of Parameters

According to the general methodology of stochastic modelling of real phenomena (see our remarks in Section 8), an important further step, after choosing an appropriate model process and determining its probabilistic characteristics, consists of relating the model parameters to measurable physical parameters. This stage can be regarded as a part of model validation, which is usually performed by comparison of the model predictions with data from repeated experiments.

The parameters of the model process in Eq. (5.1) (i.e., the statistics of a single crack increment) and the properties of the counting process $N(t)$ (e.g., the growth intensity λ_0 in the assumed Poisson process) should be estimated from empirical data. However, this directly requires a new type of fatigue experiment in which the fatigue crack growth process is regarded as a cumulative random process. In designing such fatigue experiments, care should be taken to create the possibility for identifying the underlying counting process $N(t)$, including estimation of its intensity. Also, one should prepare the experimental conditions for estimating the statistics of the random elementary crack increments. Since such experimental studies seem to be of future concern, here we show how the model parameters can be related to the traditional experimental results.

Let us assume first that the material in which the crack propagates is highly homogeneous, and that the jumps in crack growth are due only to the maxima of the loading process. In such a case, it is reasonable to approximate λ_0 as the average number of maxima of the random loading process $S(t)$ above a certain level s_0, where s_0 should be appropriately selected. (It can be taken as a fatigue limit of the material in question.) If the loading process is stationary, then λ_0 is constant and

$$\lambda_0 = \langle M_{s_0}(0, T) \rangle, \tag{5.38}$$

where the right hand-side is given by Eq. (3.17). Though approximating λ_0 in this way depends mainly on the loading process, the dependence on the material properties is partially accounted for by the suitable choice of the parameter s_0, which, in general, is associated with properties of the material (e.g., fatigue or endurance limit).

Since at least the basic information concerning the crack growth process is contained in the empirical fatigue crack growth equations, it seems to be useful to relate the parameters of a stochastic cumulative jump model with material and loading characteristics occurring in empirical crack growth equations (e.g., the Paris equation and its modifications). This can be done by use of various criteria. (See [149].)

A possible way of approximating the parameter α characterizing the average magnitude of the crack increments in the assumed exponential distribution is to use a mean-square criterion; namely, we can determine α from the condition

$$F(\alpha) = \int_0^{\bar{t}} \langle [A_P(t) - A(t, \gamma)]^2 \rangle \, dt = \min_{\alpha}, \tag{5.39}$$

where $A_P(t)$ is obtained from the Paris equation (appropriately modified when random loading is considered; see next chapter) and $A(t, \gamma)$ is from the cumulative jump model (5.1); $\bar{t}$ is the time in which $A_P(t)$ reaches critical level ξ. The criterion in Eq. (5.39) gives an effective procedure for relating parameter α to physical constants contained in $A_P(t)$. It should be noticed, however, that the value of α determined by use of Eq. (5.39) is the same for all time intervals considered. To overcome this deficiency, one can determine many values of α according to the criterion

$$F_i(\alpha) = \int_0^{t_i} \langle [A_P(t) - A(t, \gamma)]^2 \rangle \, dt = \min_{\alpha}, \tag{5.40}$$

where

$$t_i = t_{i-1} + \Delta t, \qquad \Delta t = \frac{\bar{t}}{n}, \qquad t_0 = 0, \qquad i = 1, 2, \ldots, n. \tag{5.41}$$

In this way, we obtain different values of α_i for the parameter α in different elementary time intervals on the interval $[0, \bar{t}]$.

A much weaker criterion for determination of α is as follows. We require that the global uncertainties (expressed as entropies) of the process $A(t, \gamma)$ defined in Eq. (5.1) and the random crack growth process governed by another stochastic model should be equal. The criterion is of the form

$$\int_0^{\bar{t}} H_A(t)\,dt = \int_0^{\bar{t}} H_A^*(t)\,dt, \tag{5.42}$$

where $H_A(t)$ is the one-dimensional (informational) entropy of the process $A(t, \gamma)$ defined by the model in Eq. (5.1), and $H_A^*(t)$ is the entropy of the known stochastic fatigue crack growth model; e.g., $H_A^*(t)$ can be taken as the entropy of the process obtained from a randomized Paris equation. (See next chapter.) The one-dimensional entropy of the process $A(t, \gamma)$ is defined as [77]

$$H_A(t) = -\int_0^{\infty} f_A(a;t) \ln f_A(a;t)\,da. \tag{5.43}$$

Let us assume that the structural element is subjected to random loading $S(t, \gamma)$, which is stationary and Gaussian. A modified form of the Paris–Erdogan equation (1.33) accounting for randomness of the loading is (see Chapter VI)

$$\frac{dA_P}{dt} = \mu_0 C\, g(Q)\,(K_{rms})^m, \tag{5.44}$$

where K_{rms} is twice the root mean-square of the stress intensity factor (for infinite elastic sheet, $K_{rms} = 2S_{rms}\sqrt{\pi A}$), Q is the counterpart of the stress ratio (e.g., $Q = \langle S_{\min}\rangle / \langle S_{\max}\rangle$). μ_0 is taken as the expected number of maxima of $S(t, \gamma)$ in the considered interval and is given by Eq. (3.17).

In the paper [149], it was shown that the usage of the mean-square criterion (5.39) with $A_P(t)$ determined by Eq. (5.44) with $g(Q) = 1$ leads to minimization of the following function with respect to α:

$$F(\alpha) = F_1 - F_2(\alpha) + F_3(\alpha), \tag{5.45}$$

where

$$F_1 = \int_0^{\bar{t}} A_P^2(t)\,dt,$$

$$F_2(\alpha) = 2\int_0^{\bar{t}} A_P(t)\,\langle A(t,\gamma)\rangle\,dt, \tag{5.46}$$

$$F_3(\alpha) = \int_0^{\bar{t}} \langle A^2(t)\rangle\,dt,$$

Analytical formulae for F_1 and $F_2(\alpha)$ depend on the specific numerical value of m in Eq. (5.44). For example, for $m = 4$, we obtain the following value of α:

$$\alpha = 2\bar{t}^2\left(\frac{\mu_0}{3}\bar{t}+1\right)\left[\psi(\bar{t}) - A_0\bar{t}^2\right]^{-1}, \tag{5.47}$$

where the *critical* time $\bar{t}$, defined by $A_P(\bar{t}) = \xi$, and the function $\psi(\bar{t})$ are as follows for the case considered:

$$\bar{t} = \frac{1}{C_1}\left(A_0^{-1} - \xi^{-1}\right), \tag{5.48a}$$

$$C_1 = \mu_0 C\left(2S_{rms}\sqrt{\pi}\right)^4, \tag{5.48b}$$

$$\psi(\bar{t}) = \frac{2}{C_1}\left[-\bar{t}\,\ln\xi^{-1} - \frac{1}{C_1}\left(\xi^{-1}\ln\xi^{-1} + C_1\bar{t} + A_0^{-1}\ln A_0\right)\right]. \tag{5.48c}$$

It can be shown via algebraic transformations that $\alpha > 0$. The preceding formulae give the explicit expression for the model parameter α in terms of loading characteristics S_{rms} and μ_0 and material parameters C and m, which are known from traditional experiments; α depends also on the initial crack length A_0, and because of the criterion assumed, on the critical crack length ξ, which in turn depends on the loading and material characteristics. (See Section 27.)

23.4 Numerical Illustration

To illustrate the predictions of the model discussed in this section, let us consider a common exemplary *structural element*, namely, an infinite sheet made of steel 15G2ANb. (See [136].) Experimental data (under sinusoidal loading with the stress ratio $R = 0.33$) give the following values of material parameters:

$$C = 1.03 \times 10^{-12}, \; m = 3.89, \; K_c = 52 \text{ MPa} \sqrt{\text{m}}. \tag{5.49}$$

Let us assume that a stochastic loading acting on the structural element is a Gaussian and stationary process $S(t, \gamma)$ with

$$\begin{aligned} m_S &= \langle S(t, \gamma) \rangle = \text{constant}, \\ K_S(\tau) &= S_{rms}^2 \, e^{-\beta^2 \tau^2}. \end{aligned} \tag{5.50}$$

In this case, Eq. (3.17) gives

$$\begin{aligned} \mu_0(s_0) = \frac{1}{2\pi} \{ &\beta\sqrt{6} [1 - \Phi(s_0^* \sqrt{1.5})] \\ &+ 2S_{rms} \beta \sqrt{\pi} \varphi(s_0^*) \Phi(\frac{\sqrt{2}}{2} s_0^*) \}, \end{aligned} \tag{5.51}$$

where

$$s_0^* = \frac{s_0 - m_S}{S_{rms}}. \tag{5.52}$$

Numerical values of the loading parameters are taken as follows:

$$m_S = 80 \text{ MPa}, \; S_{rms} = 55.28 \text{ MPa}. \tag{5.53}$$

The critical crack length, evaluated with use of the formula (see Section 27),

$$\xi = \frac{K_c^2}{\pi \langle S_{\max} \rangle^2}, \tag{5.54}$$

is

$$A_{cr} = \xi = 6 \times 10^{-2} \text{ m} = 6.0 \text{ cm}. \tag{5.55}$$

Assuming that $A_0 = 7 \times 10^{-3}\text{m} = 0.7$ cm, $\beta = 0.014 \text{ min}^{-1}$, and $s_0 =$ 120 MPa, we obtain $\mu(s_0) = 5.55 \text{ h}^{-1}$. The time $\bar{t}$ when $A_P(t)$ reaches ξ is 21,216 h.

The next three figures represent the results of numerical calculations for the *cumulative Poisson model* (CPM). Figure 5.1 shows the crack size distributions for the cumulative Poisson model at different time instants. The numerical values of parameter α have been obtained from the criterion in Eq. (5.40) with $n = 200$. The corresponding values of α at the three times, t_1, t_2, and t_3, are as follows: $t_1 : \alpha = \alpha_1 = 6782.041 \times 10^3$; $t_2 : \alpha = \alpha_2 = 5248.663 \times 10^3$; $t_3 : \alpha = \alpha_3 = 3608.546 \times 10^3$. It is seen that when t increases, the values of α decrease, indicating that the average value of the elementary crack increments $(1/\alpha)$ increases with time. As expected, for each specific crack size $a = \tilde{a} < \xi$, the probability $P\{A_1(t) < \tilde{a}\}$ decreases when time increases.

Figure 5.2 illustrates the dependence of the crack size distribution on the parameter s_0 characterizing the *fatiguing stress level* (n.b., the maxima of the stress process above s_0 generates crack growth). The corresponding values of μ_0 are: $\mu_0(110) = 6.47\ \mathrm{h}^{-1}$, $\mu_0(120) = 5.55\ \mathrm{h}^{-1}$, $\mu_0(130) = 4.72\ \mathrm{h}^{-1}$. Decreasing μ_0 causes an increase of $P(A_1(t) < \tilde{a})$, $(\tilde{a} \in (0, \xi - A_0))$. The curves are plotted for a fixed time instant $t = 15 \times 10^3\ \mathrm{h}$.

Figure 5.3 visualizes the effect of the correlation parameter (n.b., its inverse characterizes correlation time of the process $S(t, \gamma)$) on the crack size distribution for a fixed time instant. According to Eq. (5.51), an increasing β (i.e., decreasing of the correlation time) causes an increase of $\mu_0(s_0)$ (i.e., an increase of the infinitesimal growth intensity). As a result, we observe that $P\{A_1(t) < \tilde{a}\}$ decreases when β increases.

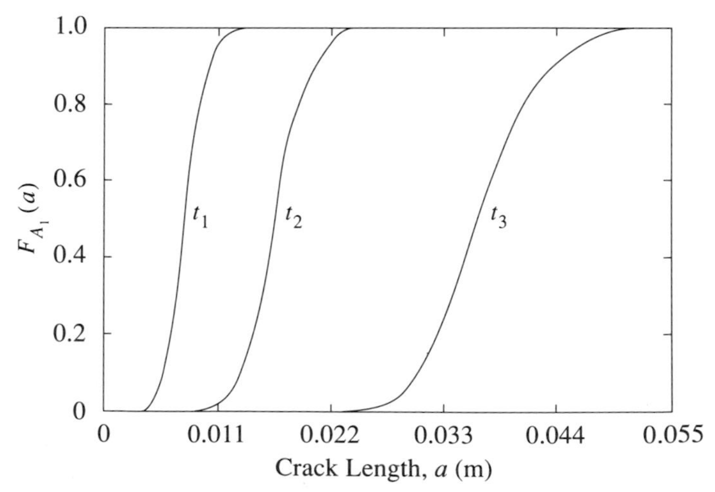

Figure 5.1. Crack size distribution (CPM) for different time instants. $\langle S_{max} \rangle = 120$, $s_0 = 120$ (MPa); $t_1 = 0$, $t_2 = 13$, $t_3 = 18$ ($\times 1000$ h).

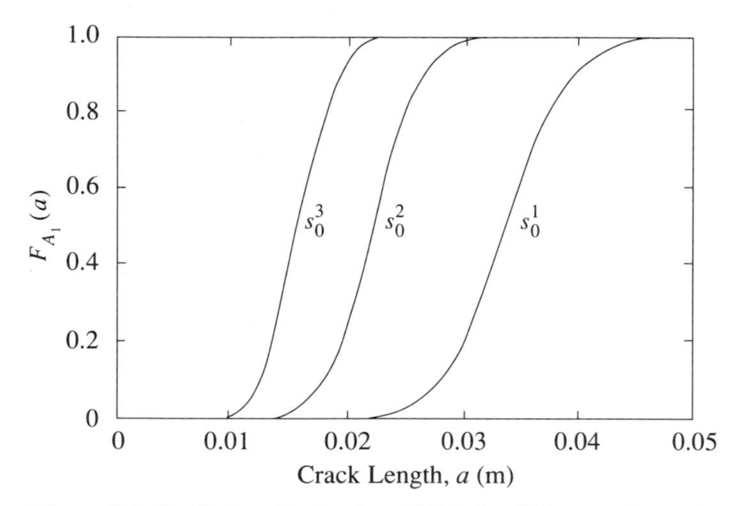

Figure 5.2. Crack size distribution (CPM) for different values of s_0. $\langle S_{max} \rangle = 120$, $s_0 = 120$ (MPa); $t_1 = 0, t_2 = 13, t_3 = 18$ ($\times 1000$ h).

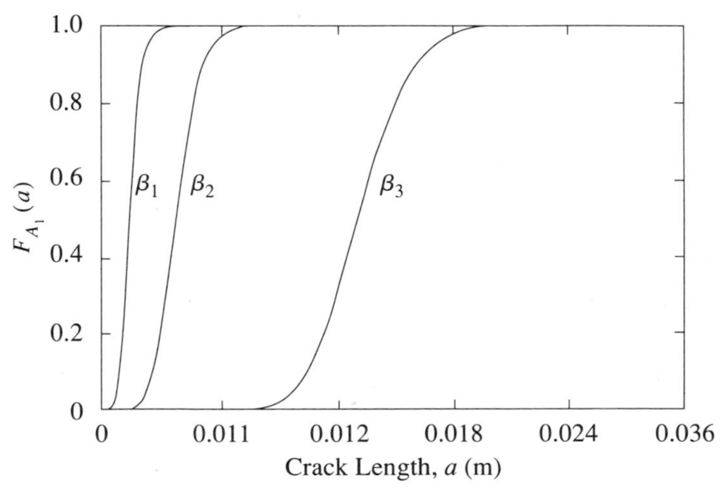

Figure 5.3. Influence of β on crack size distribution (CPM). $\langle S_{max} \rangle = 120$, $s_0 = 120$ (MPa); $\beta_1 = 1.4, \beta_2 = 3.4, \beta_3 = 8.5$ ($\times 0.01$ min^{-1}).

24. Cumulative Model with Underlying Birth Process

24.1 Basic Probability Distributions

To derive the crack size distribution, we use similar reasoning to that presented in Section 23.1. In the present case, instead of Eq. (5.6), we have

$$M(s) = e^{-\lambda^0 t} \sum_{k=1}^{\infty} [G(s)]^k (1 - e^{-\lambda^0 t})^{k-1}. \tag{5.56}$$

If the random variables $Y_i(\gamma)$ are exponentially distributed as in Eq. (5.7), then Eq. (5.8) holds. Then performing the inverse Laplace transform of Eq. (5.56),

$$\mathcal{L}^{-1}[M(s)] = \mathcal{L}^{-1}\left[\int_0^{\infty} e^{-sa} f_{A_1}(a;t)\, da\right], \tag{5.57}$$

and with the use of

$$\mathcal{L}^{-1}\left[\left(\frac{1}{s+\alpha}\right)^k\right] = \frac{a^{k-1}}{(k-1)!} e^{-\alpha a}, \tag{5.58}$$

one obtains

$$\begin{aligned} f_{A_1}(a;t) &= a e^{-\lambda^0 t - \alpha a} \sum_{k=1}^{\infty} \frac{(\alpha a)^{k-1}}{(k-1)!} (1 - e^{-\lambda^0 t})^{k-1} \\ &= \alpha\, e^{-\lambda^0 t - \alpha a} \exp[\alpha\, a (1 - e^{-\lambda^0 t})] \\ &= \alpha\, e^{-\lambda^0 t} \exp(-\alpha\, a e^{-\lambda^0 t}), \qquad a > 0, t > 0, \end{aligned} \tag{5.59}$$

and $f_{A_1}(a;t) = 0$ for $a \leq 0$. The crack size distribution is, for each fixed t, an exponential distribution, i.e.,

$$f_{A_1}(a;t) = \beta_t e^{-\beta_t a}, \qquad \beta_t = \alpha e^{-\lambda^0 t}, \tag{5.60}$$

and

$$f_A(a;t) = f_{A_1}(a - A_0;t). \tag{5.61}$$

If $A_0 = A_0(\gamma)$ is a random variable independent of $A_1(t, \gamma)$, then the remark of the previous section can be applied. (See page 130.)

Using the definitions of mean and variance and performing the appropriate integration, we obtain

$$\langle A(t, \gamma) \rangle = A_0 + \frac{e^{\lambda^0 t}}{\alpha}, \tag{5.62}$$

$$\sigma_A^2 = \frac{e^{2\lambda^0 t}}{\alpha^2}. \tag{5.63}$$

Comparison of Eq. (5.62) with Eq. (5.13) for the cumulative Poisson model (with the same values of α and $\lambda_0 = \lambda^0$) shows that the mean crack length in the birth model exceeds the corresponding curve in the Poisson model, which is consistent with the fact that the birth model assumes larger intensity of the transition of the process from state k to $(k+1)$.

The distribution function corresponding to Eqs. (5.59) and (5.61) is given by

$$F_A(a;t) = \int_0^a f_A(x;t)\,dx = 1 - \exp\left[-\alpha(a - A_0)\,e^{-\lambda^0 t}\right], \quad \alpha > A_0. \tag{5.64}$$

The distribution of the random variable T, the time at which the crack length $A(t, \gamma)$ reaches the critical value ξ, can be derived by adopting the reasoning presented in Section 23.2. Here, we shall apply Eq. (5.23) to distribution (5.64). The result is

$$F_T(t) = \exp\left[-\alpha(\xi - A_0)\,e^{-\lambda^0 t}\right]. \tag{5.65}$$

Differentiation of Eq. (5.65) with respect to t gives the following expression for the probability density function:

$$f_T(t) = \alpha^{\lambda^0}(\xi - A_0)\exp\left[-\lambda^0 t - \alpha(\xi - A_0)\,e^{-\lambda^0 t}\right]. \tag{5.66}$$

It is seen that the preceding probability density of the lifetime distribution depends on four parameters: α, λ^0, A_0, ξ. However, if we regard the quantity $\alpha(\xi - A_0) = b$ as one parameter, then Eq. (5.66) takes the form of an extreme-value type (or Gumbel) distribution,

$$f_T(t) = b\lambda^0 \exp\left[-\lambda^0 t - b e^{-\lambda^0 t}\right], \tag{5.67}$$

This result, derived from our birth cumulative model, supports the conjecture of Kozin and Bogdanoff [140], formulated on the basis of analysis of the Virkler data, that the extreme value distribution is a possible candidate for the *first passage-time* of the fatigue crack growth process.

24.2 Numerical Illustration

The discussions regarding parameter estimation of the cumulative Poisson model remain valid for the *cumulative birth model* (CBM). Both the mean-square and the entropy criteria can be used to estimate the parameters. For example, usage of the entropy criterion (5.42) requires calculation of entropy $H_A(t)$ of the distribution (5.67) and entropy $H_A^*(t)$ of the randomized Paris model. For the cumulative birth model, the integrated entropy required by criterion (5.42) is

$$\int_{t_0}^{\bar{t}} H_A(t)\,dt = (1 - \ln\alpha)\,\bar{t} + \frac{1}{2}\mu_0\bar{t}^2. \tag{5.68}$$

The calculated crack size distribution and the lifetime distribution for the cumulative birth model reported in paper [149] are shown on the next three figures. They are obtained for the same loading process as before, but the *structural element* (an infinite sheet) is made of 18G2A steel. (See [137].) The three basic material constants needed for calculations are $C = 1.95 \times 10^{-13}$, $m = 4.5$, $K_c = 138\ \text{kg/mm}^{3/2}$. It is assumed that $A_0 = 3$ mm and $\beta = 0.06\ \text{min}^{-1} = 0.001\ \text{sec}^{-1}$. When $\langle S_{\max}\rangle = s_0 = 16.3$, then $\mu_0(s_0) = 5.15\ \text{h}^{-1}$ and $\xi = 23$ mm.

Figure 5.4 shows the dependence of the crack size distribution on the *fatiguing level* s_0, whereas Fig. 5.5 illustrates the dependence on the correlation parameter β of the loading process. Figure 5.6 visualizes the effect of β on the lifetime distribution. It is seen that the change of the correlation radius of the applied stress gives an especially strong effect (within the model discussed) on the probability distributions.

25. Cumulative Jump — Correlated Model

25.1 Description of Model

The basic hypothesis of the models discussed in the previous two sections is the independence of elementary crack increments ΔA_i. However, in reality, successive damages are not independent. An accumulation of fatigue may result in a loss of resistance to further damage, so the magnitudes of successive crack jumps are, most likely, correlated to each other. Such a

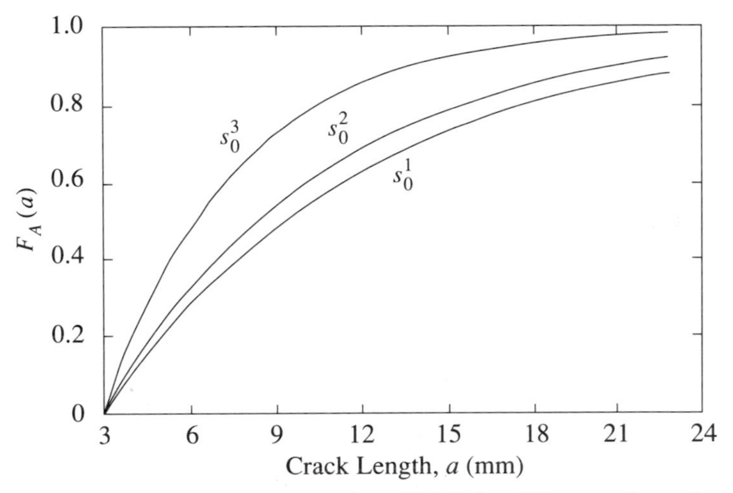

Figure 5.4. Crack Size Distribution (CBM) for different values of s_0. $s_0^1 = 13.3$, $s_0^2 = 16.3$, $s_0^3 = 19.3$ (kg/mm^2).

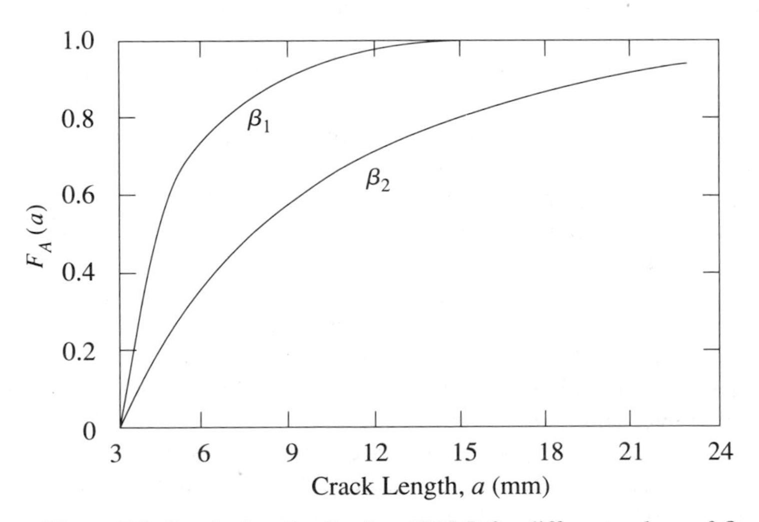

Figure 5.5. Crack size distribution (CBM) for different values of β. $s_0 = 16.3$ (kg/mm^2); $\beta_1 = 0.002$, $\beta_2 = 0.004$ (s^{-1}).

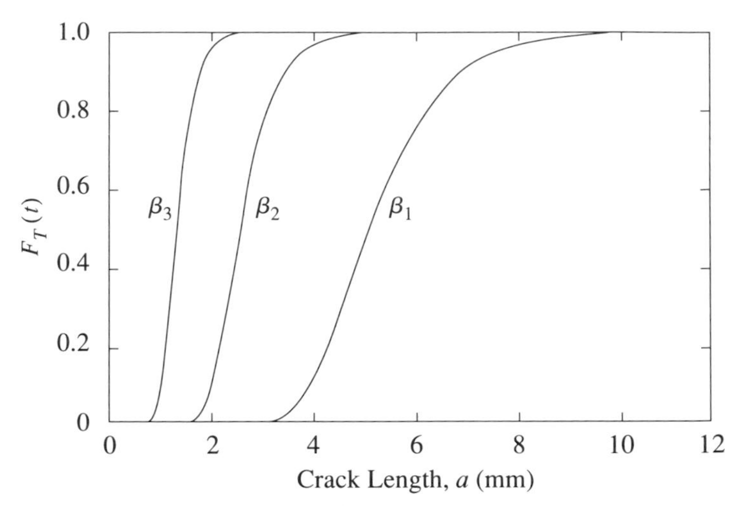

Figure 5.6. Lifetime distribution (CBM) for different values of β. $s_0 = 16.3$ (kg/mm^2); $\beta_1 = 1, \beta_2 = 2, \beta_3 = 4$ ($\times 10^{-3}$ s^{-1}) .

conjecture is supported by the observation of striation patterns for fractured materials. (See [134].) It is therefore natural to extend the cumulative model to the more general situation, where elementary crack increments are correlated random variables. Such an extension has been reported recently by Sobczyk and Trebicki [150]. Here, the basic idea will be expounded upon and illustrated.

Let us assume that fatigue crack growth is characterized by a stochastic process $A(t, \gamma)$ represented analytically by Eq. (5.1), where $N(t)$ is a homogeneous Poisson process and $Y_i(\gamma)$ are dependent random variables. Though, in general, the dependent random variables are characterized by multivariate distributions, the available information on the dependence (usually estimated from incomplete data) is most often limited to knowledge of covariances between the random variables in question. On the other hand, in many situations, marginal distributions of single variables can be prescribed on the basis of empirical data or on the physical grounds. It is therefore appropriate to construct models of multivariate distributions in terms of known one-dimensional (marginal) distributions and covariances (or correlation coefficients). A construction that we adopt here is known as the *Morgenstern model* ([142]); its generalization has been suggested by Kotz [139], whereas the conditions for validity of the model are expounded upon in the paper by Liu and Der Kiureghian [141].

Let $Y_1, Y_2, \ldots, Y_n$ be dependent random variables with given marginal densities $f_{Y_i}(y_i)$ and joint correlational moments. The joint probability density can be represented by the formula [141],

$$f(y_1, y_2, \ldots, y_n)$$
$$= \prod_i f_{Y_i}(y_i)\ \{1 + \sum_{i<j} \alpha_{ij}\,[1 - 2F_{Y_i}(y_i)]\,[1 - 2F_{Y_j}(y_j)] +$$
$$+ \sum_{i<j<k} \alpha_{ijk}\,[1 - 2F_{Y_i}(y_i)]\,[1 - 2F_{Y_j}(y_j)]\,[1 - 2F_{Y_k}(y_k)] + \ldots\},$$
$$\tag{5.69}$$

where $f_{Y_i}(y_i)$ and $F_{Y_i}(y_i)$ are the marginal density and cumulative distributions, respectively,

$$\alpha_{12\ldots k} = \frac{\rho_{12\ldots k}}{(-2)^k Q_1 Q_2 \ldots Q_k}, \tag{5.70}$$

and $\rho_{12\ldots k}$ is the k-dimensional, normalized, joint central moment,

$$\rho_{12\ldots k} = \left\langle \left[\left(\frac{Y_1 - m_1}{\sigma_1}\right)\left(\frac{Y_2 - m_2}{\sigma_2}\right) \ldots \left(\frac{Y_k - m_k}{\sigma_k}\right) \right] \right\rangle, \tag{5.71}$$

where m_i and σ_i are the mean and standard deviation of Y_i, and

$$Q_i = \int_{-\infty}^{\infty} \left(\frac{Y_i - m_i}{\sigma_i}\right) f_{Y_i}(y_i)\, F_{Y_i}(y_i)\, dy_i. \tag{5.72}$$

Of course, since the joint density (5.69) has to be nonnegative, the Morgenstern model is valid under restrictions on $\alpha_{12\ldots k}$. For instance, to assure the nonnegativeness of the bivariate density $f(y_i, y_j)$, we need $|\alpha_{ij}| < 1$, which implies the appropriate condition for $|\rho_{ij}|$.

Equation (5.69) includes, in general, all higher-order moments (5.70). However, they are seldom available in practical applications. On the other hand, using Eq. (5.69) and including many moments $\alpha_{1,2\ldots k}$ results in essential computational difficulties. Therefore, the analysis performed herein will use a truncated series preserving only binary correlation α_{ij}.

25.2 Moments of the Crack Size

25.2.1 General Formulae

We shall concentrate here on deriving the formulae for the statistical moments of the crack size represented by the jump-correlated model (5.1). This model will be represented (for convenience) in the form

$$A(t, \gamma) = A_0 + A_1(t, \gamma), \tag{5.73}$$

$$A_1(t, \gamma) = \sum_{i=1}^{N(t)} Y_i(\gamma), \tag{5.74}$$

and we will look for the statistics of the process $A_1(t, \gamma)$. Inclusion of A_0 in the analysis does not result in any additional difficulties. (See [149].)

If $N(t)$ is a homogenous Poisson process with intensity λ_0, then

$$P_n(t) = P\{N(t) = n\} = \frac{(\lambda_0 t)^n}{n!} \exp(-\lambda_0 t), \quad n = 0, 1, 2, \ldots, \tag{5.75}$$

and the mth-order moment m of $A_1(t, \gamma)$ is

$$\begin{aligned} \langle A_1^m(t, \gamma) \rangle &= \sum_{n=0}^{\infty} \langle A_1^m | N(t) = n \rangle P_n(t) \\ &= \exp(-\lambda_0 t) \sum_{n=0}^{\infty} \left\langle \left[\sum_{i=1}^{n} Y_i(\gamma) \right]^m \right\rangle \frac{(\lambda_0 t)^n}{n!} \\ &= \exp(-\lambda_0 t) \sum_{n=0}^{\infty} E_n^m \frac{(\lambda_0 t)^n}{n!}, \end{aligned} \tag{5.76}$$

where E_n^m is the mth-order conditional moment of $A_1(t, \gamma)$, that is, the mth-order moment of the sum (5.74) when the number of crack increments in the interval $(0, t]$ is fixed, $N(t) = n$. Making use of the Newton formula we have the following expression for E_n^m:

$$E_n^m = \left\langle \left[\sum_{i=1}^{n} Y_i(\gamma) \right]^m \right\rangle = \sum_{h_1 + \ldots + h_n = m} \frac{m!}{h_1! \ldots h_n!} \langle Y_1^{h_1} \ldots Y_n^{h_n} \rangle, \tag{5.77}$$

where summation is extended over all possible combinations of n elements giving the sum equal to m.

To make the preceding formulae entirely explicit, one has to evaluate the joint moments of random variables $Y_i(\gamma)$. This is accomplished by making use of the basic formula for the joint probability density given by Eq. (5.69). When only binary correlations are considered, the joint moments occurring in Eq. (5.77) are

$$\langle Y_1^{h_1} \dots Y_n^{h_n} \rangle = \prod_{i=1}^{n} \langle Y_i^{h_i} \rangle + \sum_{i<j} \alpha_{ij} \prod_{\substack{r \neq i \\ r \neq j}} \langle Y_r^{h_r} \rangle M_{h_i} M_{h_j}, \tag{5.78}$$

where

$$M_{h_r} = \int_{-\infty}^{\infty} y_r^{h_r} f_{Y_r}(y_r) \left[1 - 2F_{Y_r}(y_r) \right] dy_r. \tag{5.79}$$

Equations (5.76)–(5.79) give a general representation of the mth-order moment of the process $A_1(t, \gamma)$. For instance, the conditional moments occurring in Eq. (5.76) up to the fourth order are

$$E_n^1 = \langle \sum_{i=1}^{n} Y_i \rangle = \sum_{i=1}^{n} \langle Y_i \rangle, \tag{5.80}$$

$$E_n^2 = \sum_{i-1}^{n} \langle Y_i^2 \rangle + 2 \sum_{i<j} \langle Y_i Y_j \rangle, \tag{5.81}$$

$$E_n^3 = \sum_{i=1}^{n} \langle Y_i^3 \rangle + 3 \sum_{i<j} \langle Y_i^2 Y_j \rangle + 3 \sum_{i<j} \langle Y_i Y_j^2 \rangle + 6 \sum_{i<j<k} \langle Y_i Y_j Y_k \rangle, \tag{5.82}$$

$$\begin{aligned} E_n^4 = {} & \sum_{i=1}^{n} \langle Y_i^4 \rangle + 4 \sum_{i<j} \langle Y_i^3 Y_j \rangle + 6 \sum_{i<j} \langle Y_i^2 Y_j^2 \rangle + 4 \sum_{i<j} \langle Y_i Y_j^3 \rangle \\ & + 12 \sum_{i<j<k} \langle Y_i^2 Y_j Y_k \rangle + 12 \sum_{i<j<k} \langle Y_i Y_j^2 Y_k \rangle \\ & + 12 \sum_{i<j<k} \langle Y_i Y_j Y_k^2 \rangle + 24 \sum_{i<j<k<1} \langle Y_i Y_j Y_k Y_1 \rangle. \end{aligned} \tag{5.83}$$

The joint moments occurring in Eqs. (5.81)–(5.83) are expressed in terms of marginal densities and the binary correlations α_{ij} via Eq. (5.78). For instance,

$$\begin{aligned} \langle Y_i Y_j \rangle &= \langle Y_i Y_j \rangle + \alpha_{ij} G_1^i G_1^j, \\ \langle Y_i^2 Y_j \rangle &= \langle Y_i^2 \rangle \langle Y_j \rangle + \alpha_{ij} G_2^i G_2^j, \\ \langle Y_i Y_j Y_k \rangle &= \langle Y_i \rangle \langle Y_j \rangle \langle Y_k \rangle + \alpha_{ij} G_1^i G_1^j \langle Y_k \rangle \\ & \quad + \alpha_{ik} G_1^i G_1^k \langle Y_j \rangle + \alpha_{jk} G_1^j G_1^k \langle Y_i \rangle, \end{aligned} \tag{5.84}$$

and so on, where

$$G_p^i = \int_{-\infty}^{\infty} y_i^p f_{Y_i}(y_i) \, [1 - 2F_{Y_i}(y_i)] \, dy_i. \tag{5.85}$$

25.2.2 Exponential Distribution of Crack Jumps

Let us assume that the crack increments $Y_i(\gamma) = \Delta A_i$ are exponentially distributed with parameters η_i (i.e., the mean value of the crack increments is $1/\eta_i$). Under this hypothesis,

$$f(y_i) = \begin{cases} \eta_i \exp(-\eta_i y_i), & y_i > 0 \\ 0, & y_i \le 0. \end{cases} \tag{5.86}$$

In this case, M_{h_r} defined by Eq. (5.79), is

$$M_{h_r} = \langle Y_r^{h_r} \rangle \left(\frac{1}{2^{h_r}} - 1 \right)$$
$$\langle Y_r^{h_r} \rangle = \frac{h_r!}{(\eta_r)^{h_r}}, \tag{5.87}$$

and, according to Eq. (5.78),

$$\langle Y_1^{h_1} \dots Y_n^{h_n} \rangle = \prod_{i=1}^{n} \frac{h_i!}{\eta_i^{h_i}} \{ 1 + \sum_{i<j} \alpha_{ij} \tilde{M}_{h_i} \tilde{M}_{h_j} \}, \tag{5.88}$$

where $\tilde{M}_{h_r} = 1/2^{h_r} - 1$. The conditional moments E_n^m given by the general formula in Eq. (5.77) in our case are

$$E_n^m = \sum_{h_1 + \dots + h_n = m} m! \left[\prod_{i=1}^{n} \eta_i^{h_i} \right]^{-1} \{ 1 + \sum_{i<j} \alpha_{ij} \tilde{M}_{h_i} \tilde{M}_{h_j} \}. \tag{5.89}$$

25.2.3 Special Form of Binary Correlation

To make further results more explicit, we assume a special form of binary correlation between the crack increments, namely,

$$\alpha_{ij} = B \left[1 - \frac{|j-i|}{\max(l, j)} \right], \qquad i, j = 1, \dots, n, \quad i \ne j, B \ge 0, \tag{5.90}$$

or, from Eq. (2.2),

$$\rho_{ij} = 4Q_iQ_jB\left[1 - \frac{|j-i|}{\max\,(i,j)}\right]. \tag{5.91}$$

It is seen that, according to the physics of the problem, the correlation between the ith and jth increments is a decreasing function of the distance between jumps Y_i and Y_j. The factors Q_i and Q_j depend, according to the definition in Eq. (5.72), on the parameters of the marginal distribution. It is clear that the parameter B should be selected in such a way that the joint probability densities used for calculating the moments are nonnegative. Simple algebraic transformations lead to the appropriate range of possible values of B; for instance, if one is interested in evaluating the moments up to the third order, $B \in [0, 1]$.

Let us assume now a special form of the parameter in the exponential distribution of elementary crack increments. As the fractographical patterns of fatigue indicate, the crack jumps in its growth increase (on the average) as the number of cycles increases. Therefore, we assume here that

$$\eta_i = \frac{\eta}{i^{\nu}}, \qquad \eta > 0, \tag{5.92}$$

where ν is an integer. This implies that $\langle Y_i \rangle = i^{\nu}/\eta$ increases with the jump number i; the parameter ν can be calibrated to data.

Under the hypotheses given by Eqs. (5.90) and (5.92), Eq. (5.89) takes the form

$$E_n^m = \sum_{h_1 + \ldots + h_n = m} \frac{m!}{\eta^m}\left[\prod_{i=1}^{n} i^{\nu h_i}\right]\{1 + B\sum_{i<j} \frac{i}{j}\tilde{M}_{h_i}\tilde{M}_{h_j}\}, \tag{5.93}$$

and through algebraic transformation, the conditional moments (5.80)–(5.83) become polynomials of n of degree $m(\nu + 1)$ given by

$$E_n^m = \sum_{k=1}^{m(\nu+1)} c_k(B)\, n^k. \tag{5.94}$$

Remark: It should be noticed that Eq. (5.90) for α_{ij}, with B independent of i and j, characterizes only the positive $(B > 0)$ correlation between Y_i and Y_j; this indicates that an increase in the value of one of these random variables is related to an increase of the conditional averages of the second variable. Though such a hypothesis seems to be physically adequate and results in the possibility of obtaining Eq. (5.94), a more general analysis would require the introduction of B_{ij} instead of simply B. In this case, Eq. (5.93) would read

$$E_n^m = \sum_{h_1 + \ldots + h_n = m} \frac{m!}{\eta^m} \left[\prod_{i=1}^{n} i^{\nu h_i} \right] \{1 + \sum_{i<j} B_{ij} \frac{i}{j} \tilde{M}_{h_i} \tilde{M}_{h_j}\} . \qquad (5.94a)$$

25.2.4 Final Formulae for Moments

Let us return to the general relation given by Eq. (5.76) for the moments of random crack growth process $A_1(t, \gamma)$, and take into account the result (5.94) for the conditional moments. Equation (5.76) can be represented as

$$\langle A_1^m(t, \gamma) \rangle = \exp(-\lambda_0 t) \sum_{k=1}^{m(\nu+1)} c_k(B) \sum_{n=0}^{\infty} n^k \frac{(\lambda_0 t)^n}{n!} . \qquad (5.95)$$

The infinite series in Eq. (5.95) can be expressed in the form of the Stirling polynomials using the identity,

$$\sum_{n=0}^{\infty} n^k \frac{z^n}{n!} = \exp(z) D_k(z) , \qquad (5.96)$$

where $D_k(z)$ is a Stirling polynomial of degree k defined by the recurrence formula,

$$D_0(z) = 1,$$
$$D_k(z) = z \left[\frac{dD_{k-1}(z)}{dz} + D_{k-1}(z) \right], \qquad k = 1, 2, \ldots . \qquad (5.97)$$

Therefore,

$$\langle A_1^m(t, \gamma) \rangle = k \sum_{k=1}^{m(\nu+1)} c_k(B) D_k(\lambda_0 t) = \sum_{k=1}^{m(\nu+1)} d_k(B) (\lambda_0 t)^k, \qquad (5.98)$$

where $d_k(B)$ are obtained as a result of multiplication of $c_k(B)$ by the coefficients of $D_k(\lambda_0 t)$. For instance, when $\nu = 2$, we have the following result for the first- and second-order moments,

$$\langle A_1(t, \gamma) \rangle = \frac{\lambda_0 t}{6\eta} \{2(\lambda_0 t)^2 + 9\lambda_0 t + 6\} , \qquad (5.99)$$

$$\langle A_1^2(t,\gamma)\rangle = \frac{\lambda_0 t}{720\eta^2}\{5(\lambda_0 t)^5(16+3B) + 18(\lambda_0 t)^4(88+13B) + 345(\lambda_0 t)^3(28+3B) + 60(\lambda_0 t)^2(348+23B) + 360\lambda_0 t(38+B) + 1440\}, \tag{5.100}$$

$$\sigma_{A_1}^2 = \langle A_1^2(t,\gamma)\rangle - \{\langle A_1(t,\gamma)\rangle\}^2. \tag{5.101}$$

It should be noticed that the moments of arbitrary order m of the random crack growth process $A_1(t,\gamma)$ are functions of parameter η, which occurs in the probability distribution of elementary crack increments, the intensity of crack jumps, λ_0, and B, the parameter that characterizes the strength of the correlation between the magnitude of jumps. Symbolically, for each t, we have

$$\langle A_1^m(t,\gamma)\rangle = M_m(t;a,\lambda_0,B). \tag{5.102}$$

25.2.5 Approximation of the Crack Size Distribution

It does not seem to be possible to obtain an exact analytical formula for the probability distribution of a crack size at arbitrary time t. However, having the moments of arbitrary finite order, one can construct an approximate crack size distribution using the information contained in the moments as a basic building element. A possible approach is the use of the maximum entropy principle. (See Spencer and Bergman [152], and Sobczyk and Trebicki [151]). In the case considered herein, the maximum entropy principle can be summarized as: The best approximate probability distribution of the random process $A_1(t,\gamma)$ for each fixed t is the one that maximizes the entropy functional

$$H_t = -\int f(a;t)\ln f(a;t)\,da, \tag{5.103}$$

subject to the given moment constraints in Eq. (5.102) and the normalization condition. In [151], the maximum entropy distribution of crack size has been calculated and illustrated graphically.

25.3 Numerical Illustration

As in the models presented in Sections 23–24, the parameters of the model described here (i.e., statistics of single crack increments, the jump intensity, and the strength of the mutual correlation between elementary increments) should be inferred from the fatigue experiments that need to be designed and performed in the spirit of the model. Since such experiments have not been performed yet, the parameters of the model have been related to the physical parameters occurring in the empirical fatigue crack growth equations in a manner similar to that described in Section 23. The parameter B characterizing the strength of the correlation between crack increments should also be identified from the data. As in [150], we leave it as a free parameter and show its effect on the statistics of the crack size.

Assuming the same material and loading characteristics as in Section 23.4 and performing numerical calculations, we obtain the quantitative effect of the correlation of crack increments on the statistics of fatigue crack growth. Figure 5.7 illustrates the dependence of the variance (or standard deviation σ) of the crack size on the strength of the correlation between the crack jumps B. This effect is also shown in Fig. 5.8. These predictions agree with the features of fatigue crack growth inferred from the Virkler empirical data; the empirical standard deviation of the crack size strongly increases with the number of cycles. The maximum entropy distributions of crack size are shown on Fig. 5.9; visualized in the figure are the cumulative distribution functions $F_{A_1}(a) = P\{A_1(t, \gamma) \leq a\}$ for fixed $t = 18{,}000$ h and for several values of the correlation strength. It is seen that the effect of correlation between crack jumps on the probability distribution is significant. For instance, Fig. 5.9 indicates that $P\{A_1 < 22\text{mm}\} \approx 0$ for $B = 0$, whereas this probability is nearly 0.4 for $B = 0.2$. This indicates that the cumulative jump-correlated model leads

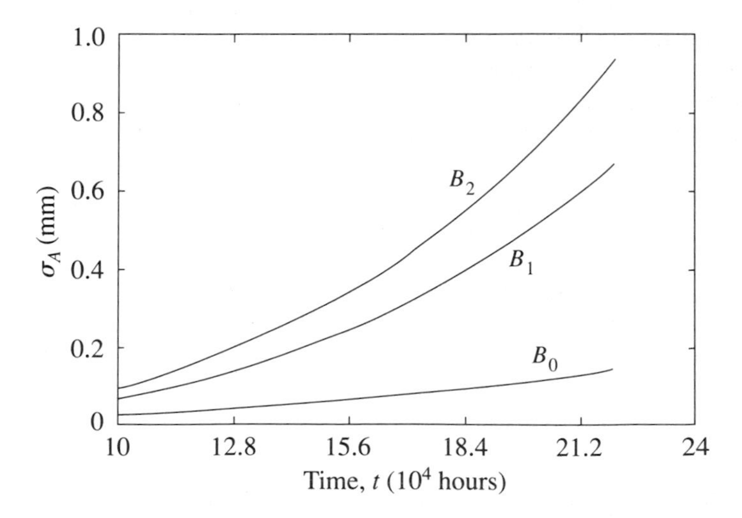

Figure 5.7. Effect of correlation strength of crack increment on standard deviation of crack size. $B_0 = 0$, $B_1 = 0.01$, $B_2 = 0.02$.

to longer specimen lifetimes as compared to predictions of the cumulative jump-uncorrelated model. This and other features of the model require further analysis, including inference from experimental data.

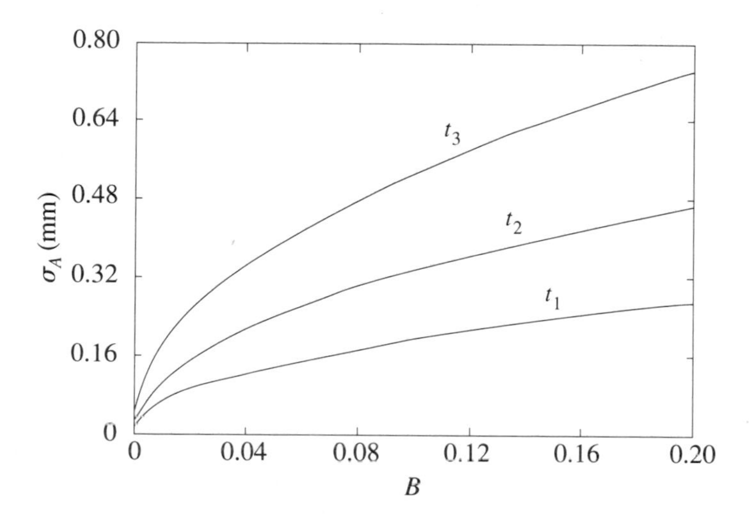

Figure 5.8. Standard deviation of crack size versus correlation strength B. $t_1 = 15$, $t_2 = 18$, $t_3 = 21$; $\times 10^3$ hours.

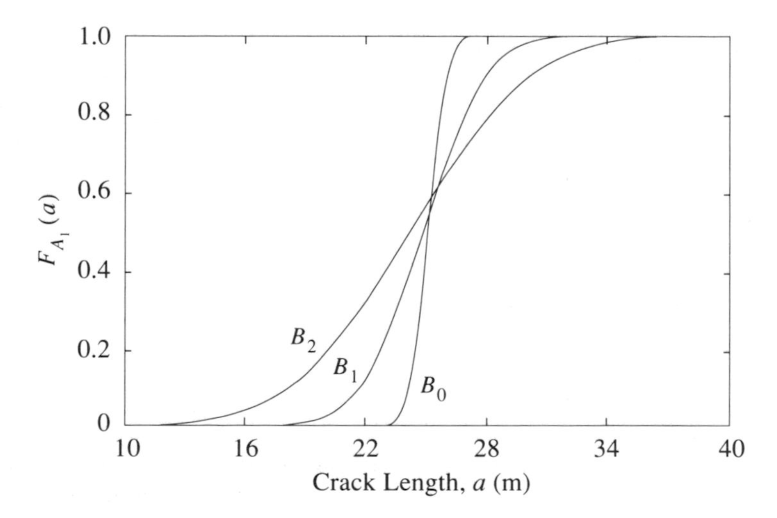

Figure 5.9. Effect of B on the maximum entropy distributions of crack size for fixed t. $t = 18{,}000$ hours, $B_0 = 0$, $B_1 = 0.05$, $B_2 = 0.2$.

26. Random Fatigue Crack Growth with Retardation

26.1 Crack Growth Retardation due to Overloads

As we have said in Section 4.3, one of the important effects associated with crack growth under variable-amplitude loading is the decrease in growth rate, termed *crack retardation*, which normally follows a high overload. As pointed out by many researchers, this retardation can have a significant influence on the fatigue life of a structure. If the retarding effect of a peak overload on the crack growth is neglected, the prediction of the material lifetime is usually very conservative. Accurate predictions of the fatigue life will hardly become possible before the physics of the peak overload mechanisms are better clarified. According to the existing findings, the retardation is a physically very complicated phenomenon that is affected by a wide range of variables associated with loading, metallurgical properties, environment, etc. It is difficult to separate the contributions of each of these variables.

Although the retardation phenomenon is quite complex, the existing experimental results along with proposed physical and theoretical models provide an essential contribution to the understanding of the influence of overloading on the fatigue crack growth in real materials. Taking into consideration the observations reported in recent papers on fatigue crack growth retardation, we can draw the following important conclusions [131], [135], [143], [146], [153].

1. Positive overloads introduce significant crack growth retardations. In general, however, longer retardations are generated by [131]: (i) increasing the magnitude of the overload, (ii) repeating the overload during the crack growth, and (iii) applications of blocks of overloads instead of single overloads.
2. Crack growth retardation does not occur immediately after an overload.
3. Comparison of crack growth for program loading and for random loading has revealed sequence effects; these effects were not restricted to crack growth rates but were also observed in the structure of the fracture surface.
4. Compressive overloads have a relatively small effect on crack growth. However, when a tensile overload is followed immediately by a compressive overload, the crack growth retardation is greatly diminished. If the compressive overload precedes the tensile overload, the reduction of the retardation is much smaller. Most of the reported prediction methods are based on the assumption that only rising tensile load ranges cause growth.

5. Retardation effects dominate over acceleration effects. Acceleration in fatigue crack growth is generally significant at high stress levels. Since high loads are expected to occur infrequently and are usually of short duration, the influence of crack acceleration on fatigue life prediction may be neglected [154]. Retardation of the fatigue crack growth, on the other hand, can be particularly large at the lower stress levels relevant under normal operating conditions [154], [155].
6. Retardation is crack size-dependent; for a fixed overload ratio, the number of cycles with retarded crack growth usually increases with increasing crack size.
7. Retardation of the crack growth is affected by a broad range of metallurgical, loading, and environmental variables, and it is very difficult to separate the contributions of each source. In particular, the retardation depends on the ductility of the material; if the ductility of an alloy is controlled by heat treatment, a lower yield strength will produce longer periods of retardation. As loading factors are concerned, experiments show a permanent reduction in growth rate with increasing stress biaxiality.

These basic features of the retardation phenomenon indicate its complexity. Nevertheless, numerous investigators have made efforts to find appropriate physical explanations for the retarded crack growth and to treat retardation quantitatively. Frequently cited models are the effective plastic zone model proposed by Wheeler and the effective crack closure model of Elber. However, there is no general agreement concerning the ability of these and other recently proposed models with respect to satisfactory prediction of fatigue crack growth [144], [145].

Stouffer and Williams [153] characterize the situation in strong words: "Although the literature contains a number of attempts to model this phenomenon through manipulation of the constants and stress intensity factors in the equations of Paris–Erdogan, little appears to have been done in the effort to develop a completely rational analysis of the problem."

It is likely that one reason that the existing models of retarded crack growth are not satisfactory is that these models are deterministic, while the fatigue crack growth phenomenon shows strongly random features. In addition, most of the reported theoretical descriptions of retardation are based on data fitting techniques that tend to hide the behavior of the phenomenon. It is therefore of interest to look at the crack growth process with retardation from a more general point of view that takes into account the inherent randomness of the phenomenon.

Various attempts to construct a probabilistic model for fatigue crack growth with retardation have been recently reported in the literature. (See Arone [130], Ditlevsen and Sobczyk [133], Sobczyk [148], and Winterstein and Veers [156].) Here, we shall characterize briefly the model presented by Ditlevsen and Sobczyk. In this model, fatigue crack growth is characterized by a random cumulative process, and retardation due to overloads is accounted for by modification of the infinitesimal growth intensity of the birth process.

26.2 Periodic Loading with Separated Overloads

Let us consider first the crack growth due to periodic loading with a single peak overload. This process (assumed to be random due to material heterogeneity) is characterized by the cumulative model process (5.1), where $N(t)$ is a birth process. It is assumed that the crack growth under homogeneous cyclic loading is governed by process (5.1), with the birth process $N(t)$ having a time-independent intensity $\lambda_k = \lambda^0 k$, $k = 1, 2, \ldots$, $\lambda^0 > 0$. To describe crack retardation due to overload, it is postulated that

$$\lambda_k = \lambda_k(t) = \lambda^0 [k - \lambda_{OL}(t, k; t_1)], \qquad 0 < \lambda_{OL} < \lambda^0, \tag{5.104}$$

where $\lambda_{OL}(t, k; t_1)$ characterizes a retarding effect of an overloading that has occurred at time t_1.

As indicated in Section 26.1, the retarding intensity λ_{OL} depends on several variables, such as the magnitude of the overload (overload ratio ρ), the crack size, the stress biaxiality, the ductility of the material, etc. To obtain a practicable model, λ_{OL} is restricted to have the form,

$$\lambda_{OL}(t, k; t_1) = \lambda_{OL}(t; t_1) f(k), \tag{5.105}$$

where $f(k)$ isolates the influence of the crack size on the retardation. The reported observations show that the amount of retardation increases with increasing crack size. However, the existing experimental results do not seem to be sufficient to indicate an appropriate functional form of $f(k)$. For mathematical convenience, it is simply assumed here that $f(k) = k$. Thus, Eq. (5.105) takes the form,

$$\lambda_{OL}(t, k; t_1) = k\lambda_{OL}(t; t_1). \tag{5.106}$$

The function $\lambda_{OL}(t; t_1)$ is given the form,

$$\lambda_{OL}(t; t_1) = \mu(t, \zeta)\, (t - t_1 - \theta)^{\beta}\, e^{-\alpha(t - t_1 - \theta)} H(t - t_1 - \theta), \tag{5.107}$$

where $H(\cdot)$ is the Heaviside unit step function given by

$$H(t) = \begin{cases} 1 & \text{for } 0 \leq t \\ 0 & \text{otherwise,} \end{cases} \tag{5.108}$$

and $\mu(t, \zeta)$ is a suitably bounded retardation magnitude function that may depend on time t and a collection of relevant variables denoted by ζ (e.g., the overload ratio, the stress biaxiality, etc.). The integral ratio (i.e., the geometrical center point),

$$\frac{\int_0^\infty t^{1-\beta} e^{-\alpha t} dt}{\int_0^\infty t^{\beta} e^{-\alpha t} dt} = \frac{\beta + 1}{\alpha}, \tag{5.109}$$

may be interpreted as a characteristic retardation time. The parameter θ is introduced to reflect a delay of the start of the retardation, which may be modelled to be more or less abrupt by suitable choice of the parameter β.

This model characterizes a random crack growth process as a cumulative process (5.1), with $N(t)$ being a birth process of the growth intensity given by Eqs. (5.104)–(5.108).

The physical features of the phenomenon enter into the model through the parameters ζ, α, β, and θ. The interpretation of some of these parameters in terms of experimental data will be discussed in Section 26.4.

No problems arise in extending the model to the slightly more general case where the crack growth is generated by periodic loading with N overloads of different magnitude that are sufficiently separated to assure that the retardation period after each overload is effectively over before the next overload occurs.

Although there may be some sequence effects, their influence will be neglected here; sequence effects are often considered secondary as compared to the stress magnitude effects [132]. In this case, the growth intensity is postulated to have the form,

$$\begin{aligned} \lambda_k &= \lambda_k(t; t_1, t_2, \ldots, t_N) \\ &= \lambda^0 k \prod_{i=1}^{N} [1 - \mu(t, \zeta_i)(t - t_i - \theta_i)^{\beta_i}] \\ &\quad \times e^{-\alpha_i(t - t_i - \theta_i)} H(t - t_i - \theta_i) . \end{aligned} \tag{5.110}$$

A situation that is often investigated experimentally concerns blocks of overloads. It has been observed that when blocks of overloads are applied instead of a single overload, a longer duration of the retardation is obtained. If the overload blocks are well separated, it seems reasonable to characterize this case by a retarding intensity of the form as in Eq. (5.110), where t_i is a characteristic instant for the ith overload block and $\mu(t, \zeta_i)$ is replaced by $\bar{\mu}(t, \zeta_i) = B_i \mu(t, \zeta_i)$. The quantity $B_i > 1$ describes the block effect. Additionally, the retardation time $(\beta_i + 1)/\alpha_i$ should be appropriately characterized to be relevant to the block overload (e.g., instead of α_i, one can take $\bar{\alpha}_i = \delta_i \alpha_i$, where δ_i describes the block effect on the retardation time, $\delta_i < 1$).

26.3 Periodic Loading with Interaction of Overloads

If the time intervals between overloadings are not long enough to assure that no essential interaction of the retardation effects occurs, the situation is much more complicated. The single retardation effects appear less clearly in the observed crack growth curves. The effects of interaction of overloadings may, in fact, be positive (retardation) or negative (accelerated growth).

The existing experimental results give no clear indication of how to quantitatively characterize the interaction effects of overloads. Many mechanisms seem to contribute to these effects, and it is difficult to recognize and separate them. It is observed that application of a high overload, after a previous overload has become effective, may considerably reduce the crack growth retardation. Within the cumulative birth model considered herein, a possible way of modelling the crack growth under periodic loading with irregularly but deterministically occurring overloads of the same size is to define the growth intensity as

$$\lambda_k(t) = \lambda_{int}(t)\, \lambda_k, \tag{5.111}$$

where $\lambda_{int}(t)$ is an interaction factor. It may be written as

$$\lambda_{int}(t) = 1 + \psi(t; \tau_i, \tau_{i-1}, \ldots), \tag{5.112}$$

where $\psi(t; \tau_i, \tau_{i-1}, \ldots)$ is a function of present time t and the time instants $\ldots < \tau_{i-1} < \tau_i < t$ of occurrence of the overloads. The simplest interaction model is obtained by letting ψ be a function $\psi(t, \tau_i, \tau_{i-1})$ solely of t, τ_i, and τ_{i-1}, decreasing in $t - \tau_{i-1}$ such that $\psi(t, \tau_i, -\infty) = 0$. A suggestion is

$$\lambda_{int}(t) = 1 + \xi(\tau_i)\, e^{-\kappa(t - \tau_{i-1})}, \tag{5.113}$$

where κ is some *memory* parameter. The factor $\xi(\tau_i)$ quantifies the interaction effect after the latest overload.

If the overloads have different sizes, the factor ξ in Eq. (5.113) may depend in a complicated way on the specific sequence of overloads. To suggest anything about that dependency seems far beyond the present knowledge. At most, it may be suggested to use Eq. (5.113), with ξ being an appropriate random variable $\xi(\gamma)$ or, perhaps, a stochastic process that characterizes an overall effect at time t of the irregularity of the size and the occurrence of the overloads. In such a case, the process $N(t)$ in the model process (5.1) becomes a doubly stochastic birth process.

26.4 Crack Size Distribution

To obtain the probability distribution of crack size at arbitrary time t, it is convenient to evaluate the characteristic function of the process $A(t)$ represented by Eq. (5.1), where $N(t)$ is a birth process with time-varying intensity, i.e., $\lambda_k = k\lambda(t)$. For periodic loading with a single overload, the growth intensity is

$$\lambda(t) = \lambda^0 [1 - \mu(t, \zeta)(t - t_1 - \theta)^{\beta} e^{-\alpha(t - t_1 - \theta)} H(t - t_1 - \theta)]. \quad (5.114)$$

In the case of multiple separated overloads, $\lambda(t)$ is defined by Eq. (5.110), whereas in the more general situation of interacting overloads, it is defined by Eq. (5.111). The system of differential equations for the probability $P_k(t)$ that $N(t) = k$ is as follows:

$$\frac{dP_k(t)}{dt} = -\lambda(t) k P_k(t) + \lambda(t)(k-1) P_{k-1}(t), \qquad k > 1,$$
$$\frac{dP_1(t)}{dt} = -\lambda(t) P_1(t). \quad (5.115)$$

This system of equations can be reduced to a system with constant coefficients of the same form (in which $\lambda = 1$) by setting $\lambda(t) = d\eta/dt$, $\eta(0) = 0$, or

$$\eta(t) = \int_0^t \lambda(s)\, ds. \quad (5.116)$$

The solution for $P_k(t)$ is

$$P_k(t) = e^{-\eta(t)} [1 - e^{-\eta(t)}]^{k-1} \qquad k = 1, 2 \quad (5.117)$$

Assume that $\mu(t, \zeta) = \mu(\zeta)$; that is, the retardation magnitude function depends only on the variables ζ that cause retarded growth. Essentially, ζ includes the overloading ratio $\rho = S_{\max, OL}/S_{\max}$. Furthermore, assume that these variables are time-independent. For a single overload, the function $\eta(t)$ then becomes

$$\eta(t) = \lambda^0 t, \quad t < t_1, \tag{5.118a}$$

$$\begin{aligned}\eta(t) &= \lambda^0 \left[t - \mu(\zeta) \int_0^{t - t_1 - \theta} s^\beta e^{-\alpha s} ds \right] \\ &= \lambda^0 \{ t - \mu(\zeta) \frac{\beta!}{\alpha^{\beta+1}} [1 - e^{-\alpha(t - t_1 - \theta)}] \sum_{i=0}^{\beta} \frac{(\alpha t)^i}{i!} \}, \quad t > t_1.\end{aligned} \tag{5.118b}$$

In the case of multiple separated overloads, a direct generalization is obtained by using Eq. (5.110).

For periodic loading with interaction of the overloads, the growth intensity of the underlying birth process may be a stochastic process as suggested in the preceding. In such a case, Eq. (5.117) gives the conditional probability, $P\{N(t) = k | \eta(t)\}$. Of course,

$$\begin{aligned}P_k(t) &= \langle e^{-\eta(t, \gamma)} [1 - e^{-\eta(t, \gamma)}]^{k-1} \rangle \\ &= \int_0^\infty e^{-\eta_t} (1 - e^{\eta_t})^{k-1} dF_t(\eta),\end{aligned} \tag{5.119}$$

where $F_t(\eta)$ is the one-dimensional distribution of $\eta(t, \gamma)$, and

$$\begin{aligned}\varphi_{A(t)}(u) &= \langle \exp(iuA_t(\gamma)) \rangle = \left\langle \exp\left(iu \left[A_0 + \sum_{j=1}^{N(t)} Y_j(\gamma) \right] \right) \right\rangle \\ &= \sum_{k=1}^{\infty} \langle \exp(iuA_t(\gamma)) | N(t) = k \rangle P_k(t) \\ &= \sum_{k=1}^{\infty} \left\langle \exp\left(iu \left[A_0 + \sum_{j=1}^{k} Y_j(\gamma) \right] \right) \right\rangle.\end{aligned} \tag{5.120}$$

Due to the independence of the random variables $Y_j(\gamma)$ and the independence of $Y_j(\gamma)$ and A_0 (which is assumed here to be a random variable), we have

$$\begin{aligned}\varphi_{A(t)}(u) &= e^{-\eta(t)} \sum_{k=1}^{\infty} \varphi_{A_0}(u) [\varphi_Y(u)]^k (1 - e^{-\eta(t)})^{k-1} \\ &= e^{-\eta(t)} \varphi_{A_0}(u) \varphi_Y(u) \frac{1}{1 - \varphi_Y(u) [1 - e^{-\eta(t)}]},\end{aligned} \tag{5.121}$$

since

$$\sum_{k=1}^{\infty} x^k = \frac{1}{1-x}, \qquad |x| < 1. \tag{5.122}$$

The probability density of $A(t)$ is the Fourier transform of $\varphi_{A(t)}(u)$ given by Eq. (5.121) and it can be evaluated in particular cases. The mean and variance may be calculated by use of the first- and second-order derivatives of Eq. (5.121) at $u = 0$, or they may be evaluated directly.

26.5 Relation to the Empirical Models

The model described previously needs appropriate empirical verification by comparison with data. In the case of reasonable qualitative agreement between the behavior of the model and the observations, the characteristics of the model should be estimated from the data from repeated experiments on fatigue crack growth. In designing the appropriate fatigue experiments and in elaborating the results, the methods of statistical inference for jump random processes should be adopted.

As in previous sections, it is possible to relate the parameters occurring in the retardation model just described to the existing empirical predictions. For instance, the intensity λ^0 of the birth process can be related to the quantities occurring in the Paris–Erdogan empirical equation. In [133], such estimation of λ^0 is made by comparison of the solution of Paris–Erdogan equation and the average value of the crack size predicted by the cumulative jump model.

As the parameters directly connected with retardation are concerned, the identification creates more serious problems. This is primarily due to the fact that there is no commonly accepted empirical model of the retardation phenomenon. If we accept the Wheeler model (i.e., Eq. (1.51)), then the estimation of the retardation time (i.e., Eq. (5.109)) can be performed as follows.

To simplify the reasoning, let us assume that β in the expression for λ_{0L} is equal to zero. In this case, $1/\alpha$ is the retardation time, so it can be assumed to be

$$\alpha^{-1} = \frac{1}{\omega} N_{ret}, \tag{5.123}$$

where N_{ret} is the number of cycles of retarded growth (due to overloading). In the case when the retardation factor C_p in the Wheeler model (i.e., Eqs. (1.51) and (1.52)) is defined as (see Broek and Smith [132])

$$C_p \approx \left(\frac{r_p}{r_{po}}\right)^p, \tag{5.124}$$

where r_p is the size of the plastic zone at crack length A corresponding to constant amplitude loading, and r_{po} is the size of the plastic zone caused by the overload, we can obtain an explicit expression for N_{ret}. Indeed, since (see Eq. (1.23))

$$r_p = \frac{K_{max}^2}{cS_y^2}, \quad r_{po} = \frac{K_{max,\,OL}^2}{cS_y^2}, \tag{5.125}$$

we obtain

$$C_p \approx \rho^{-2p}, \tag{5.126}$$

where the overloading ratio ρ is defined as

$$\rho = \frac{S_{max,\,OL}}{S_{max}}. \tag{5.127}$$

The Wheeler equation (1.51) yields (we assume, for simplicity, $R = 0$, i.e., $K_{min} = 0$)

$$\begin{aligned} \left(\frac{dA}{dN}\right)_{OL} &= \rho^{-2p} C (\Delta K)^m \\ &= \rho^{-2p} C K_{max}^m. \end{aligned} \tag{5.128}$$

Since

$$\Delta r = r_{po} - r_p = \frac{1}{cS_y^2}\left[K_{max,\,OL}^2 - K_{max}^2\right], \tag{5.129}$$

the number of cycles of retarded growth is

$$N_{ret} = \frac{\Delta r_p}{\left(\frac{dA}{dN}\right)_{ret}} = \frac{K_{max,OL}^2 - K_{max}^2}{cS_y^2\, \rho^{-2p} C K_{max}^m}$$
$$= \frac{\rho^{2p}}{cS_y^2} \frac{\rho^2\text{-}1}{C S_{max}^{m-2} (\pi A_*)^{(m-2)/2}}, \tag{5.130}$$

where A_* is the crack length at which overload occurred. This estimate of N_{ret} agrees with empirical observations showing that higher overloads produce longer retardation periods. In the general case, when C_p is given by Eq. (1.52), the lower and upper bounds for N_{ret} can be obtained by integration of the differential equation defined by the Wheeler model. (See [133].)

26.6 Cumulative Model with Exponential Decay

Another possible approach to modelling fatigue crack growth with retardation can be based on the following reasoning. (See Sobczyk [148].) Let us assume that in the absence of overloading, the crack is growing according to the constant-amplitude fatigue growth equation, and the crack size at time t is characterized by $A_{CA}(t)$; $A_{CA}(t)$ can be regarded as deterministic or random. The retardation due to an overload can be accounted for by assuming that the net crack size is considered.

Let us assume that n reported overloads occur in the time instants $t_1, t_2, \ldots, t_n$; then the following representation of the crack size at arbitrary time t can be assumed:

$$A(t, \gamma) = A_{CA}(t) - A_{OL}(t, \gamma; t_1, t_2, \ldots, t_n), \tag{5.131}$$

where

$$A_{OL}(t, \gamma; t_1, t_2, \ldots, t_n) = \sum_{i=1}^{n} X_i(\zeta, \gamma)\, e^{-\alpha_i (t - t_i)} H_i(t - t_i), \tag{5.132}$$

and $X_i(\zeta, \gamma)$ is a nonnegative random factor characterizing the magnitude of retardation at time t due to the ith overload, which occurred at $t = t_i$ and depends on the collection of variables denoted symbolically by ζ; α_i is the retardation parameter characterizing the time of retarded growth after each overload; $H_i(t - t_i) = H(t - t_i) - H(t - t_i - \tau_i)$; and τ_i is the retardation time associated with the ith overload.

The model process, Eqs. (5.131) and (5.132), is relatively simple, so its basic statistical characteristics can be effectively evaluated. If we assume that the retarding factors $X_i(\zeta, \gamma)$ are mutually independent random variables, then the model leads to an effective analytical expression for the characteristic function of the crack size; namely,

$$\begin{aligned}\varphi_A(u) &= \langle \exp[iuA(t)] \rangle = \varphi_{A_{CA}}(u)\, \varphi_{A_{OL}}(-u) \\ &= \varphi_{A_{CA}}(u) \prod_{i=1}^{n} \varphi_{X_i}[-u e^{-\alpha_i(t-t_i)} H_i(t-t_i)]\,.\end{aligned} \tag{5.133}$$

Of course, the mean and variance of the crack size (for the case considered) are

$$\langle A(t,\gamma) \rangle = \langle A_{CA}(t) \rangle - \sum_{i=1}^{n} \langle X_i(\zeta,\gamma) \rangle\, e^{-\alpha_i(t-t_i)} H_i(t-t_i)\,, \tag{5.134}$$

$$\sigma_A^2(t) = \sigma_{A_{CA}}^2(t) + \sum_{i=1}^{n} \sigma_{X_i}^2\, e^{-2\alpha_i(t-t_i)} H_i(t-t_i)\,. \tag{5.135}$$

The model process $A(t,\gamma)$ represented by Eqs. (5.131) and (5.132) requires the estimation of the statistics of the random retarding factors $X_i(\zeta,\gamma)$ from empirical data. Since the magnitude of retardation strongly depends on the overloading ratio ρ, the random variables $X_i(\zeta,\gamma)$ can be constructed as the suitable functions of ρ. The retardation parameters α_i are inversely proportional to the retardation times following each overload and can be related to the empirical data via the Wheeler model (similar to the previous section).

The cumulative model with exponential decay can be extended to describe fatigue crack growth with retardation more adequately. In particular, we mean the crack growth with randomly occurring overloads and the effects of their interaction. It seems that a proper stochastic model process for the crack growth with randomly occurring (separated) overloads can be represented in the form

$$A(t,\gamma) = \begin{cases} A_{CA}(t)\,, & N(t) = 0 \\ A_{CA}(t) - \sum_{i=1}^{N(t)} X_i(\zeta,\gamma)\, e^{-\alpha_i(t-t_i)} H_i(t-t_i)\,, & N(t) > 0, \end{cases} \tag{5.136}$$

where $N(t)$ is a counting random process characterizing a number of overloads in the interval $[0, t]$, and t_i are random instances of occurrence of overloads.

The analysis of such models, and analogous ones, will be a future concern of the authors.

27. Generalization and Remarks

The models presented in the previous sections are based on representation of the crack size at time t in the form of the sum of a countable number of crack increments (jumps), and elementary fatigue damage was identified with the magnitude of the crack increment. However, it is clear that fatigue damage can be related to the elementary crack increments in a more complex way. Also, the incremental damage may depend both on the magnitude of the elementary crack increment and on the damage level already existing in the material. Such a more general formulation can be accomplished by use of the counting integral with respect to a random Poisson measure. (See [77] and [147].)

Let us denote a random process characterizing the amount of fatigue damage at time t by $D_F(t, \gamma)$, and its initial value at $t = t_0$ by D_F^0. An increment of the process at the infinitesimal time interval Δt can be represented as

$$\Delta D_F(t, \gamma) = \int_I h(t, y, \gamma) N(\Delta t, dy), \tag{5.137}$$

where h is a suitable random function of time and variable y belonging to the interval $I \subseteq R_1^+$. The variable y denotes here a random magnitude of elementary crack jumps occurring at time t; in other words (and more generally), the variable y can be regarded as a mark that is used to characterize an event (the crack increment) at time t. The function $h(t, y, \gamma)$ characterizes the amount of *elementary* damage caused by a crack increment occurring at time t and having a mark (e.g., magnitude) y. It is assumed that $N(t, y)$ is a Poisson random function defined on $[t_0, \infty) \times R_1^+$; it characterizes the number of events that occurred in the time interval $[t_0, t]$ and had marks y.

It is clear that the overall fatigue damage due to events (e.g., due to elementary crack jumps) that occurred in the time interval $[t_0, t]$ with marks belonging to I is

$$D_F(t, \gamma) = \int_{t_0}^{t}\int_{I} h(s, y, \gamma)\, N(ds, dy) + D_F^0. \tag{5.138}$$

The preceding integral also can be represented in the form of the sum [147]

$$D_F(t, \gamma) = D_F^0 + \sum_{k=1}^{N_t} h_{s_k}(Y_k), \qquad N_t \ge 1, \tag{5.139}$$

where

$$N_t = \int_{t_0}^{t}\int_{I} N(ds, dy) \tag{5.140}$$

characterizes a random number of jumps occurring in the interval $[t_0, t]$ with magnitudes y belonging to interval I.

A more general model process for fatigue damage (when the incremental damage depends also on the damage level existing in the material just before the damaging jump) can be represented as

$$D_F(t, \gamma) = D_F^0 + \int_{t_0}^{t}\int_{I} h(s, D_F(s^-), y)\, N(ds, dy). \tag{5.141}$$

The damage increases by the amount characterized by the function $h(s, D_F(s^-), y)$, where y is the magnitude of the crack increment at time s, and $D_F(s^-)$ is the damage level just before the jump occurs. As it is seen, Eq. (5.141) constitutes a stochastic integral equation with respect to a Poisson measure. (See [147].)

Chapter VI

Random Fatigue Crack Growth: Differential Equation Models

28. Introductory Remarks

In the construction of the models presented in the two preceding chapters, emphasis was focused on the probabilistic mechanisms of fatigue accumulation, whereas the true physical properties of the fatigue process are introduced via appropriate parameter estimation on the basis of empirical fatigue data. Although such an approach to modelling real phenomena is widely accepted, it seems nevertheless to be important to expound on the physics of the fatigue process in a more explicit way, e.g., by directly relating the random factors and processes provoking fatigue and the mechanisms of fatigue crack growth.

A possible approach to such modelling is to make use of the empirical fracture mechanism equations and introduce into them appropriate quantities characterizing random effects. First attempts in this direction were aimed at incorporating the characteristics of a random loading into the empirical growth models deduced from fatigue data under constant-amplitude loads. Such modified fatigue crack growth equations, being still deterministic, include, for example, the root mean square of random applied stress instead of stress range for constant-amplitude loading. (See Swanson [192], Barsom [159], and Dover and Hibbert [166].) The main features of this approach will be presented in the next section.

Another more satisfactory approach to modelling random fatigue with the use of empirical crack growth equations consists of explicit introduction of random quantities into the empirical equations (i.e., randomization of empirically motivated equations). Since most of the empirical models for fatigue crack growth (see Section 4.2) have the form of differential equations, a randomization leads to stochastic differential equations for fatigue crack growth. Such an idea (known also in other fields, e.g., in physics and population dynamics) has recently attracted many researchers and turned out to be both fruitful and promising. (See Bolotin [160], [161], Ditlevsen [162], Ditlevsen and Olesen [163], Dolinski [164], [165], Guers and Rackwitz [171], Kozin and Bogdanoff [174], Madsen [178], Lin and Yang [176], [177], Ortiz and Kiremidjian [182], [183], Sobczyk [127], [187], Spencer and co-workers [190], [191], [194], Tsurui and Ishikawa [195], Tsurui et al. [196], and Yang et al. [200].)

Although the randomization approach seems to be attractive and promising, it should also be realized that it raises some subtle questions of a methodical nature. For example: (i) How should one randomize deterministic empirical equations to obtain the most adequate stochastic models? (ii) How should the randomized equations and their solutions be interpreted? (iii) If we wish to obtain adequate and effective models, what kind of approximations and hypotheses in building stochastic equation models are admissible?

The last question seems to be of special importance. The empirical differential equations for fatigue crack growth are evidently nonlinear and complicated. Therefore, the corresponding stochastic differential equations are usually not amenable to effective analytical treatment, and the probability distributions of the solution needed for reliability prediction can only be obtained under some restrictive hypotheses or numerically. So far, most of the stochastic differential equation models for fatigue crack growth have resulted from randomization of Paris–Erdogan-type equations. It is clear that stochastic models based on other empirical equations may also bring insight into the random fatigue process.

In this chapter, the basic features of stochastic differential equation models will be described and illustrated. Since the subject is now in the process of intense development, many problems will be left open for future research work.

29. Modified Crack Growth Equation for Random Loading

Let us assume that a structural element is subjected to a time-varying random loading and that the material is linearly elastic with known stress intensity factor K. (See Section 4.1.) Let $S(t, \gamma)$ be a stochastic process characterizing the random applied stress; we assume that the mean $m_S(t)$ and the correlation function $K_S(t_1, t_2)$ are known; in particular, the root mean square S_{rms} is: $S_{rms}(t) = [K_S(t, t)]^{1/2}$. In many situations, the process $S(t, \gamma)$ may be regarded as stationary. Then $m_S(t) = m_S$ = constant, $K_S(t_1, t_2) = K_S(t_2 - t_1)$, and $S_{rms}(t) = S_{rms}$ = constant.

To adopt the empirical crack growth equations for the case considered, appropriate modifications are necessary. The basic quantity controlling fatigue crack growth under cyclic loading is the stress intensity factor range ΔK. Various quantities associated with the random stress process $S(t, \gamma)$ have been proposed to play the role of the stress range ΔS. Most early studies suggest that ΔS should be replaced by double S_{rms}. It seems, however, that a more adequate quantity could be the mean range of the random process $S(t, \gamma)$, that is, according to Sections 13.3 and 13.4, the quantity

$$S_{mr} = \langle \Delta S \rangle = \langle S_{\max} \rangle - \langle S_{\min} \rangle, \tag{6.1}$$

where $S_{\max} = m_S + Z$, $S_{\min} = m_S - Z$, and Z is a random height of peaks with probability distribution (3.23). Therefore, the mean range is $S_{mr} = 2\langle Z \rangle$. After integration, we have the result,

$$S_{mr} = 2S_{rms}\sqrt{\frac{\pi}{2}(1 - \epsilon^2)}, \tag{6.2}$$

where ϵ is the spectral width parameter defined by Eq. (3.19). If the process $S(t, \gamma)$ is a narrow-band one, then $\epsilon \to 0$ and $S_{mr} = 2S_{rms}\sqrt{\pi/2} = \sqrt{2\pi}S_{rms}$.

Another modification concerns the stress ratio R. It is suggested in [166] that for the case of random loading, it is proper to use a new ratio, Q, to indicate the mean stress content; this ratio is defined as $Q = \langle S \rangle / S_{rms}$. The most direct modification of the stress ratio (which for periodic constant-amplitude loading is defined by Eq. (1.2)) would be as follows:

$$Q = \frac{\langle S_{\min} \rangle}{\langle S_{\max} \rangle} = \frac{m_S - \langle Z \rangle}{m_S + \langle Z \rangle}. \tag{6.3}$$

A further departure from the constant-amplitude approach is concerned with the concept of a cycle (or frequency) that plays the basic role in the conventional fatigue analysis; the empirical crack growth rate is usually expressed as dA/dN, where N is a number of cycles. In the case of sinusoidal loading, there is, of course, a simple relation to the temporal description, since $N = \omega t$, where ω is a frequency. When the load is randomly varying in time, the definition of a cycle is not straightforward and not unique. In the case of random loading, a relation between number of cycles N and time t is random, i.e.,

$$N = N(t, \gamma), \tag{6.4}$$

where $N(t, \gamma)$ is a random counting process depending on the definition of a *cycle*. For a narrow-band random process, the definition is straightforward, since such a process has a single predominant frequency. Therefore, the equivalent frequency ω_e defined by Eq. (3.16) can be used. For broad-band random loadings, $N(t, \gamma)$ is usually taken as the number of local maxima of the process $S(t, \gamma)$ in the time interval $[t_0, t]$. Then the mean of $N(t, \gamma)$ characterizing the average peak frequency is given by Eq. (3.17).

Taking into account the aforementioned modifications, the averaged crack growth under random loading can be represented in the form of the following differential equation,

$$\frac{dA}{dt} = \omega_S F(Q, \Delta K_{mr}), \tag{6.5}$$

where K_{mr} is the mean stress intensity range, and ω_S is the appropriate *frequency* of $S(t, \gamma)$, e.g., the number of peaks of $S(t, \gamma)$ in unit time.

For the case when the *structural element* is an infinite sheet, we have

$$\Delta K_{mr} = S_{mr}\sqrt{\pi A}, \tag{6.6}$$

and the modified equation (6.5) of Paris–Erdogan form is

$$\begin{aligned} \frac{dA}{dt} &= C\omega_S(\Delta K_{mr})^m \\ &= C\omega_S \pi^{m/2} S_{mr}^m A^{m/2} \\ &= C_1 S_{mr}^m A^{m/2}, \qquad C_1 = C\omega_S \pi^{m/2}. \end{aligned} \tag{6.7}$$

The solution of the preceding equation satisfying the initial condition $A(t_0) = A_0$ is given by Eq. (1.42), in which C_1 is as Eq. (6.7) and ΔS is replaced by S_{mr}.

It is seen that in the approach described in this section, the random loading is replaced by an *equivalent* deterministic cyclic loading, whose frequency and amplitude are expressed by averaged characteristics of the original random loading process. The analysis of fatigue crack growth along this line is rather simple and can be appealing in practice; however, there is still not sufficient information on the reliability of such a prediction of fatigue. In the model described, the averaged characteristics do not reflect specific features of the loading spectra and the associated properties of the probability distributions.

30. Random Variable Models

30.1 General Approach

As indicated in Section 4.2, fatigue crack growth can be characterized by the following nonlinear differential equation:

$$\frac{dA}{dt} = F[A, \Delta S, C, \ldots], \tag{6.8}$$

where A and ΔS are the crack size and stress range, respectively, whereas C denotes symbolically the material parameters affecting crack growth.

A possible approach to modelling random fatigue with the use of empirical crack growth equations of the general form of Eq. (6.8) is a randomization of Eq. (6.8) by introducing appropriate random variables. They can characterize the scatter in stress range, material properties, or environmental factors.

In general, Eq. (6.8), when randomized, can be written as

$$\begin{aligned} \frac{dA(t)}{dt} &= F[A(t), V_1(\gamma), \ldots, V_m(\gamma), t], \\ A(t_0) &= A_0(\gamma), \end{aligned} \tag{6.9}$$

where $V_1(\gamma), \ldots, V_m(\gamma)$ symbolize the appropriate random variables introduced into the empirical fatigue crack growth equation, and $A_0(\gamma)$ is in general a random variable characterizing the initial crack size.

If Eq. (6.9) has a solution

$$A(t) = g_t[A_0, V_1, \ldots, V_m], \tag{6.10}$$

which, for each t, possesses a one-to-one inverse transformation ($m+1$ equations)

$$\begin{aligned} A_0 &= h_t[A, V_1, \ldots, V_m], \\ V_1 &= V_1, \ldots, V_m = V_m, \end{aligned} \tag{6.11}$$

then the joint probability density of the solution $A(t)$ and random variables $V_1, \ldots, V_m$ is given by the known formula,

$$f_t(a, v_1, \ldots, v_m) = f_0[h_t(a, v_1, \ldots, v_m), v_1, \ldots, v_m]\,|J|, \tag{6.12}$$

where $f_0(a_0, v_1, \ldots, v_m)$ is a joint probability density of the initial condition and random variables $V_1, \ldots, V_m$, and J is the Jacobian of the inverse transformation, i.e.,

$$J = \begin{vmatrix} \frac{\partial h_t}{\partial a} & \frac{\partial h_t}{\partial v_1} & \cdots & \frac{\partial h_t}{\partial v_m} \\ 0 & 1 & \ddots & 0 \\ \vdots & \vdots & & \vdots \\ 0 & 0 & \cdots & 1 \end{vmatrix} . \tag{6.13}$$

It can be shown (see Sobczyk [77]) that the density $f(a, v_1, \ldots, v_m; t)$, which is the solution of Eq. (6.9), satisfies the following first-order partial differential equation known as the Liouville equation:

$$\frac{\partial f}{\partial t} + \sum_{i=0}^{m} \frac{\partial}{\partial y_i}(F_i f) = 0, \tag{6.14}$$

where $F_0 = F$ and $F_1 = V_1, \ldots, F_m = V_m$. The initial condition associated with Eq. (6.14) is

$$f(a, v_1, \ldots, v_m; t_0) = f_0(a_0, v_1, \ldots, v_m). \tag{6.15}$$

In the situations when Eq. (6.9) cannot be solved analytically to obtain solution (6.10), the Liouville equation (6.14) can be used to obtain a numerical solution.

Of course, when the solution given by Eq. (6.10) is found in analytical form, it can be used directly to obtain the moments of the process $A(t)$; namely,

$$\langle A^k(t) \rangle = \int \ldots \int g_t^k(a_0, v_1, \ldots, v_m) f_0(a_0, v_1, \ldots, v_m)\, da_0 dv_1 \ldots dv_m. \tag{6.16}$$

What class of random processes is generated by the randomization of empirical fatigue crack growth equations via introduction of random variables? It can be shown (starting from the definition of a Markov process) that the process $A(t)$ is Markovian. This is confirmed by the Liouville equation (6.14), which is a special case of the Fokker–Planck–Kolmogorov equation (i.e., Eq. (2.108)) for diffusion Markov processes.

The general description given in the preceding indicates how the basic statistical characteristics of the solution of differential equations with random variables can be evaluated. However, randomization of deterministic empirical equations creates some additional questions. For example, how can we randomize the empirical equations?

30.2 Specific Random Variable Models

Let us consider a special case of general Eq. (6.8), namely, the well- known Paris–Erdogan equation (1.31) given by

$$\frac{dA}{dt} = C(\Delta K)^m, \tag{6.17}$$

where C and m are empirical constants and ΔK is the stress intensity factor range. According to Eq. (1.22), the quantity ΔK can be represented, for a wide class of practical situations, in the form,

$$\Delta K \approx B \Delta S A^{1/2}, \tag{6.18}$$

where B is a parameter related to the geometry of a specimen. Thus, Eq. (6.17) can be written as

$$\frac{dA}{dt} = C(B \Delta S A^{1/2})^m, \qquad A(t_0) = A_0, \tag{6.19}$$

or

$$\frac{dA}{dt} = C_0 A^{m/2}, \tag{6.20}$$

where $C_0 = C(B\Delta S)^m$ now characterizes all geometric, stress, and material parameters. If C_0 is not a function of A or t, Eq. (6.19) is deterministic with the following solution:

$$A(t) = \left[A_0^{(2-m)/2} - \frac{2-m}{2} C_0 (t - t_0)\right]^{2/(2-m)}, \qquad m \neq 2. \tag{6.21}$$

In the case of $m = 2$, we obtain an exponential function of t. (See Eq. (1.43).) Equation (6.19) and its solution (6.21) contain three parameters: A_0, C_0, and m. Therefore, the most direct way to randomization in this case consists of replacing these constants (all or some of them) with random variables.

Let us assume first that only the initial condition A_0 is random with probability distribution function $f_{A_0}(a_0)$, i.e., $A_0 = A_0(\gamma)$. Since the initial crack size includes much uncertainty (including the meaning of the *initial crack size* itself), choosing A_0 as a random variable seems to be very natural. In such a case, Eq. (6.21) gives $A(t)$ as a nonlinear transformation of $A_0(\gamma)$. Thus, the probability distribution of $A(t)$, for each fixed t, is a special case of Eq. (6.12); the explicit relation is given by

$$\begin{aligned} f_A(a;t) &= f_{A_0}\left([a^{-\beta} - \beta C_0 t]^{\frac{-1}{\beta}}\right)\frac{d}{da}[a^{-\beta} - \beta C_0 t]^{\frac{-1}{\beta}} \\ &= f_{A_0}\left([a^{-\beta} - \beta C_0 t]^{\frac{-1}{\beta}}\right)[a^{-\beta} - \beta C_0 t]^{-1/\beta - 1} a^{-m/2}, \end{aligned} \tag{6.22}$$

where $\beta = (m-2)/2$. Therefore, all statistical characteristics of the process $A(t)$ can be obtained if the probability density of the initial crack size is given. It should be noticed, however, that though the randomization of the initial condition allows us to account for the statistical scatter in the initial state of the fatigue crack growth process, it does not account for the variability of the realizations of crack growth from specimen to specimen. Indeed, as can be seen from Eq. (6.21), sample functions of $A(t)$ are basically the same curves simply shifted up or down the a axis. (See Fig. 6.1.)

To account for the material variability of specimens, let us randomize Eq. (6.19) by regarding C_0 as a random variable, i.e., $C_0 = C_0(\gamma)$. Since C_0 should be nonnegative, it is convenient to assume that $C_0(\gamma)$ is lognormally distributed (i.e., the *lognormal random variable model*; see Yang et al. [200], [201],[202]). We get the following stochastic equation:

$$\frac{dA}{dt} = C_0(\gamma) A^{m/2}, \qquad A(t_0) = A_0. \tag{6.23}$$

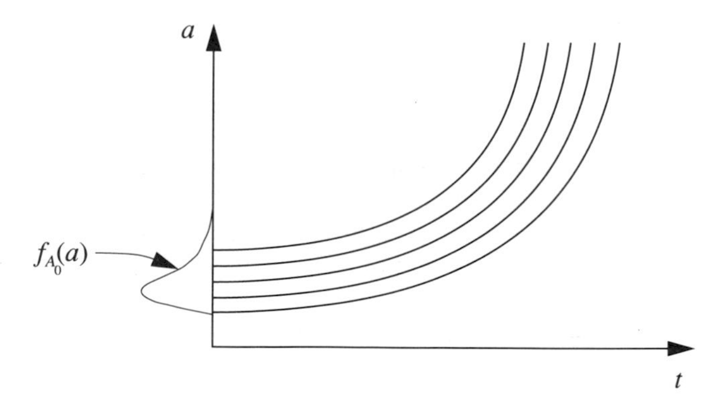

Figure 6.1. Sample functions of crack growth with random initial crack size.

The solution of Eq. (6.23) for each realization of the random variable $C_0(\gamma)$ has the form of Eq. (6.21), or, equivalently,

$$A(t) = \frac{A_0}{[1 - C_0(\gamma)\,\beta\, A_0^{\beta}(t - t_0)]^{1/\beta}}, \qquad \beta = \frac{m-2}{2}. \tag{6.24}$$

Since $C_0(\gamma)$ is lognormally distributed, its cumulative probability distribution is

$$F_{C_0}(c_0) = P\{C_0 \le c_0\} = \Phi\left(\frac{\ln c_0 - \mu_Z}{\sigma_Z}\right), \tag{6.25}$$

where μ_Z and σ_Z are the mean and standard deviation of a normal random variable $Z = \ln C_0$, respectively, and Φ is the standard normal distribution function.

The probability distribution of the crack size $A(t)$ at each time instant t can be obtained from that of $C_0(\gamma)$ through the transformation given by Eq. (6.24). The result is

$$F_A(a;t) = P\{A(t) \le a\} = \begin{cases} 0, & a \le A_0 \\ \Phi\left(\dfrac{\ln \eta_t - \mu_Z}{\sigma_Z}\right), & a > A_0, \end{cases} \tag{6.26}$$

where $\eta_t = (A_0^{-\beta} - a^{-\beta}) / \beta(t - t_0)$.

Let $T(a_*)$ be a random variable denoting the time to reach any given crack size a_*. Then $T(a_*)$ can be obtained from Eq. (6.24) by setting $A(t) = a_*$ and $t = T(a_*)$, respectively, i.e.,

$$T(a_*) = \frac{1}{\beta C_0(\gamma)} (A_0^{-\beta} - a_*^{-\beta}) + t_0. \tag{6.27}$$

The probability distribution of $T(a_*)$ is given by the formula

$$F_T(t) = P\{T(a_*) \le t\} = \begin{cases} 0, & a^* \le A_0 \\ 1 - \Phi\left(\dfrac{\ln \eta_t^* - \mu_Z}{\sigma_Z}\right), & a^* > A_0, \end{cases} \tag{6.28}$$

where $\eta_t^* = (A_0^{-\beta} - a_*^{-\beta}) / \beta(t - t_0)$.

Using the preceding results, Yang et al. [202] estimated the random number of flight hours to reach various crack sizes a_*. Comparison with the experimental test results (for fastener holes) showed a good correlation with theoretical predictions. Analysis of other random variable models can be found in a number of papers. (See Madsen [178], Akita and Ichikawa [157], and Provan [186].)

30.3 Relation to Deterministic Models

In Section 28, one of the questions that we raised was: How should one interpret the randomized equations and their solutions? Related questions are: What is a possible interpretation of the traditional Paris–Erdogan equation in the case of statistically dispersed data? Does the Paris–Erdogan equation represent a mean equation (as a result of averaging over possible scattered results), or does it describe the crack growth in each specimen for a given (fixed) value of the constant C_0?

These two interpretations should not be confused. To see the problem, let us notice that the mean value $\langle A(t) \rangle$ of the solution of stochastic equation (6.23) does *not* satisfy the equation

$$\frac{d\langle A\rangle}{dt} = \tilde{C}_0 \langle A\rangle^{\tilde{m}/2}, \tag{6.29}$$

where $\tilde{C}_0$ and $\tilde{m}$ are some fixed deterministic constants. This is obviously due to the fact that

$$\langle C_0(\gamma) A^{m/2} \rangle \neq \langle C_0(\gamma) \rangle \langle A^{m/2} \rangle \neq \langle C_0(\gamma) \rangle \langle A \rangle^{m/2}. \tag{6.30}$$

A question that arises in this context is whether there exists any other quantity (aside from the crack size) that, when averaged, satisfies the equation of the same form as its sample function equation.

Let $T(a)$ be a time in which crack size reaches the length a. For the randomized model (6.23), $T(a)$ is a random function that, for each γ, can be obtained from the solution given in Eq. (6.21). Indeed,

$$a^{(2-m)/2} = A_0^{(2-m)/2} + \frac{2-m}{2} C_0(\gamma) T \tag{6.31}$$

and

$$T(a) = \frac{2}{2-m} \frac{1}{C_0(\gamma)} \left[a^{(2-m)/2} - A_0^{(2-m)/2} \right]. \tag{6.32}$$

Differentiation of Eq. (6.32) with respect to a gives

$$\frac{dT(a)}{da} = \frac{1}{C_0(\gamma)} a^{-m/2}. \qquad (6.33)$$

Since the dependent variable does not occur on the right-hand side of Eq. (6.33), the equation for $\langle T(a) \rangle$ is given by

$$\frac{d\langle T(a) \rangle}{da} = \langle \frac{1}{C_0(\gamma)} \rangle \, a^{-m/2}. \qquad (6.34)$$

Therefore, Eq. (6.34) for sample functions has the same form as the equation for mean. This observation can be interpreted so that in the study of random fatigue crack growth, instead of randomizing constant C_0 in the Paris–Erdogan equation, one might use Eq. (6.34), since the mean version and the sample version have the same form and, therefore, the equation with $T(a)$ as the dependent variable can be interpreted as a sample or as a mean equation. (See also Kozin and Bogdanoff [174].) Although such a similarity of sample and mean versions is a nice feature of a model, it does not seem to be necessary in building stochastic models for fatigue. Of primary importance in specific situations are the proper interpretations of both deterministic empirical fatigue crack growth equations and their randomized versions.

In this context, it is worth noting that introducing random elements into nonlinear differential equations may cause the solution — in contrast to the deterministic case — to be defined on a time interval of random duration. This is clearly seen from the solution shown in Eq. (6.24); for example, for $m = 4$ and $A_0 = 1$, this solution, for each $\gamma \in \Gamma$, is finite only for time instants $t \in [0, C_0^{-1}(\gamma)]$. This feature of the solution indicates that special care should be taken to select truly adequate probability distributions for the parameters over which randomization is carried out.

31. Paris Equation with White Noise

31.1 General Properties

Randomization of empirical crack growth equations by random variables can be regarded as a particular case of randomization by a stochastic process; a random variable can simply be viewed as a *constant*, or totally correlated, random process. More general randomization, which at the same time is physically more capable, consists of introducing an appropriate stochastic process $X(t, \gamma)$ into the empirical equation. This means that instead of the deterministic equation (6.8), we consider the following stochastic differential equation:

$$\frac{dA}{dt} = F[A, \Delta S, C, \ldots] X(t, \gamma), \tag{6.35}$$

where $X(t, \gamma)$ is a nonnegative random process representing the combined effects of unknown random factors causing scatter in crack growth rates as a function of time. In fact, the model given by Eq. (6.35) is flexible enough so that $X(t, \gamma)$ can be used to better describe the effect of variations in stress ratio that often are assumed to occur multiplicatively.

There are two basic requirements concerning the process $X(t, \gamma)$: (i) It should characterize the random interactions affecting crack growth as adequately as possible, and (ii) it should have features that allow an effective treatment of Eq. (6.35), including determination of the probability distribution of a crack size.

It is not easy to find a satisfactory compromise between these two postulates. Usually, adequate characterization of real (nonnegative) random noises makes the analysis of the corresponding stochastic equations very involved, whereas models allowing nice analytical treatment are often not adequate enough. As a matter of fact, the existing efforts in modelling of fatigue crack growth by stochastic differential equations oscillate between these two objectives.

To recognize the basic properties of the randomized crack growth equation, we shall start from a special case where the random disturbance $X(t, \gamma)$ is a Gaussian white noise. Such a process is nonphysical, but turns out to be a very useful idealization. An additional deficiency of this process in our context stems from its Gaussian nature, which, theoretically, takes negative values with positive probability. This deficiency, however, can be significantly diminished if we assume that random fluctuations characterized by the white noise are small and that the deterministic positive part of the equation dominates the behavior of the process.

Therefore, let us assume that (see Sobczyk [188])

$$\begin{aligned} X(t, \gamma) &= m_X + \sqrt{2D}\,\xi(t, \gamma), \\ \langle \xi(t_1, \gamma)\, \xi(t_2, \gamma) \rangle &= \delta(t_2 - t_1), \end{aligned} \tag{6.36}$$

where m_X is the constant mean value of $X(t, \gamma)$ and $2D$ is the intensity of the fluctuations of $X(t, \gamma)$, which are considered to be small as compared to m_X. Hence, Eq. (6.35) takes the form,

$$\frac{dA}{dt} = F[A, \Delta S, C, \ldots]\, m_X + F[A, \Delta S, C, \ldots] \sqrt{2D}\,\xi(t, \gamma). \tag{6.37}$$

Because the quantities ΔS, C, m_X are assumed to be given, Eq. (6.37) can be written as

$$\frac{dA}{dt} = m(A) + \sigma(A)\,\xi(t, \gamma), \tag{6.38}$$

where $m(A) = F[A, \Delta S, C, \ldots]\, m_X$, $\sigma(A) = F[A, \Delta S, C, \ldots]\sqrt{2D}$.

Although Eq. (6.38) with $\xi(t, \gamma)$ being a white noise looks like a differential equation, it is really only a formal record of symbols, since $\xi(t, \gamma)$ is not a stochastic process in the usual sense. Thus, Eq. (6.38) should be treated as a *pre-equation* that can be turned into a meaningful equation via introduction of an appropriate interpretation. (See [77].) There are two common interpretations of *pre-equation* (6.38) when turning it into a meaningful stochastic differential equation defining the process $A(t, \gamma)$: the Itô and Stratonovich interpretations. Here, we shall adopt the Stratonovich interpretation. This means that Eq. (6.38) is understood as the following Stratonovich equation:

$$\text{(S)} \qquad \begin{aligned} dA(t) &= m(A)\,dt + \sigma(A)\,dW(t), \\ A(t_0) &= A_0, \end{aligned} \tag{6.39}$$

where $W(t)$ is the Wiener process starting from $t = t_0$ and $W(t_0) = 0$ almost surely. The preceding equation is equivalent to the following Itô stochastic differential equation [77],

$$\text{(I)} \qquad \begin{aligned} dA(t) &= m^*(A)\;dt + \sigma(A)\,dW(t), \\ A(t_0) &= A_0, \end{aligned} \tag{6.40}$$

where

$$m^*(A) = m(A) + \frac{1}{2}\sigma(A)\frac{\partial \sigma}{\partial A}. \tag{6.41}$$

It should be noted that in the Stratonovich interpretation (which in modelling of physical problems is very natural), the deterministic drift term $m^*(A)$ in the corresponding Itô equation differs from the *macroscopic* deterministic term in Eq. (6.38). In addition to the original deterministic term $m(A)$ occurring in empirical law, the Itô equation (6.40) contains a second component that comes from the existence of random multiplicative noise in the Stratonovich interpretation of *pre-equation* (6.38). So, the noise modifies the *macroscopic* empirical behavior of the crack; this is analogous to the phenomenon that in statistical physics is known as the noise-induced transition.

If the coefficients of the Itô stochastic differential equation (6.40) satisfy the appropriate conditions [77], then the solution $A(t)$ is continuous with probability 1, and it is a diffusion Markov process with the following drift and diffusion coefficients:

$$\tilde{a}(A) = m^*(A), \qquad \tilde{b}(A) = \sigma^2(A). \tag{6.42}$$

Therefore, the density $p(A, t| A_0, t_0)$ of the transition probability satisfies the Fokker–Planck–Kolmogorov equation (see Eq. (2.119)),

$$\frac{\partial p}{\partial t} + \frac{\partial}{\partial A}[m^*(A)p] - \frac{1}{2}\frac{\partial^2}{\partial A^2}[\sigma^2(A)p] = 0. \tag{6.43}$$

Hence, a randomization of the empirical crack growth equations by multiplicative white noise leads directly to the Markov diffusion model discussed in Section 19. Due to the direct randomization, the infinitesimal characteristics of the diffusion process (i.e., drift and diffusion coefficients) have been related to the parameters occurring in the fracture mechanics empirical equations.

If the coefficients in the basic Itô stochastic differential equation (6.40) grow more quickly (as functions of A) than linearly, then, in general, the solution describes the crack growth process only up to a random time $\tau(\gamma)$, which is called an *explosion time.* Up to $\tau(\gamma)$, the solution of Eq. (6.40) is characterized by a diffusion Markov process with coefficients (6.42). The random time $\tau(\gamma)$ is the time in which the process $A(t)$ *explodes* to infinity; it can be regarded as the point at which unstable crack growth occurs, resulting in ultimate fatigue failure of a specimen.

31.2 Crack Size Distribution

To illustrate the ideas described in the previous section, we consider a specific empirical fatigue crack growth equation, namely, the Paris–Erdogan equation (1.31) in the form of Eq. (6.19). Thus, Eq. (6.37) takes the form,

$$\frac{dA}{dt} = c_1 A^\rho + c_2 A^\rho \xi(t, \gamma), \tag{6.44}$$

where $\rho = m/2$, $c_1 = C_0 m_X$, and $c_2 = C_0\sqrt{2D}$. This indicates that the coefficients of the stochastic Itô equation (6.40) are

$$m(A) = c_1 A^\rho, \qquad \sigma(A) = c_2 A^\rho, \qquad m^*(A) = c_1 A^\rho + \frac{1}{2}\rho c_2^2 A^{2\rho - 1}. \tag{6.45}$$

The analysis of the Itô stochastic differential equation (6.40) with coefficients (6.45) depends on the numerical values of ρ. (See [188].) If $\rho = 1$, the equation is linear and can be solved directly with the result

$$A(t) = A_0 \exp\{c_1(t-t_0) + c_2[W(t) - W(t_0)]\}. \tag{6.46}$$

Since the increments of the Wiener process $W(t)$ have a Gaussian distribution, the preceding solution indicates that for each fixed t (when A_0 is assumed deterministic), $A(t)$ has a lognormal distribution.

If $\rho > 1$, the probability distribution of a crack size can be obtained by solving the partial differential equation (6.43) or by solving the stochastic differential equation (6.40) with coefficients (6.45). Here, we will deal with the stochastic equation directly. Equation (6.40) with coefficients (6.45) takes the form

$$\begin{aligned} dA(t) &= \left[c_1 A^{\rho}(t) + \frac{1}{2}\rho c_2^2 A^{2\rho-1}(t)\right] dt + c_2 A^{\rho}(t)\, dW(t), \\ A(t_0) &= A_0. \end{aligned} \tag{6.47}$$

The preceding equation should be considered on the finite interval $[A_0, A_{cr}]$, so at the ends of this interval, the appropriate boundary conditions should be posed. Since the main objective of this section is to expound on the basic features of stochastic differential models, we show here the analytical treatment of Eq. (6.47) without boundary conditions. Of course, to capture all important physical features of the fatigue crack growth process, more adequate stochastic differential models (i.e., with boundary conditions) are necessary.

Let us transform the process $A(t)$ into a simpler process $Y(t)$ via the following transformation:

$$Y_t = \varphi(A_t) = \frac{1}{A_t^{\rho-1}}, \qquad t \geq t_0. \tag{6.48}$$

By virtue of the Itô formula [77], we get the following equation for the process $Y(t)$:

$$\begin{aligned} dY(t) &= -(\rho-1)c_1 dt - (\rho-1)c_2 dW(t), \\ Y(t_0) &= Y_0. \end{aligned} \tag{6.49}$$

Therefore, the process $Y(t)$ is governed by a simple linear stochastic differential equation with analytical solution given by

$$Y(t) = Y_0 - (\rho-1)c_1(t-t_0) - (\rho-1)c_2[W(t) - W(t_0)]. \tag{6.50}$$

To facilitate the transformation back to the original variable, let us introduce

$$\Delta_t = \frac{1}{A_0^{\rho-1}} - \frac{1}{A_t^{\rho-1}} \\ = (\rho-1)\,c_1\,(t-t_0) + (\rho-1)\,c_2\,[W(t) - W(t_0)]\,, \tag{6.51}$$

where $A_0 = a_0$ is the deterministic initial crack size. Δ_t is normally distributed with the following mean and variance:

$$\langle\Delta_t\rangle = (\rho-1)\,c_1\,(t-t_0)\,, \\ \sigma^2_{\Delta_t} = (\rho-1)^2 c_2^2\,(t-t_0)\,. \tag{6.52}$$

Therefore, the probability density of the crack size at time t is given by

$$f_A\,(a_t;t) = \frac{\rho-1}{a_t^{\rho}}\,\frac{1}{[2\pi\sigma_{\Delta_t}]^{1/2}}\exp\left[-\frac{(a_0^{1-\rho} - a_t^{1-\rho} - \langle\Delta_t\rangle)^2}{2\sigma_{\Delta_t}}\right]. \tag{6.53}$$

From the preceding probability density, one can obtain other distributions of interest in fatigue life prediction. (See Eqs. (4.38)–(4.40).)

It should be noted that the distribution given in Eq. (6.53) describes the probability structure of the diffusion Markov process $A(t)$, which is an approximation of the physical process of crack growth. In fact, the distribution given by Eq. (6.53) also admits those values of $A(t)$ that are smaller than a_0; otherwise, the total probability would not be equal to unity (n.b., we solved the governing stochastic equation without boundary conditions). The error that arises from using Eq. (6.53) in the analysis of real fatigue crack growth behavior (i.e., $A(t) \geq a_0$) is greatest when t is near t_0, and decreases as $t - t_0$ increases. Therefore, in predicting long fatigue lives, the error can be considered negligible. (See also Lin and Yang [177].) To better see this feature of the solution, notice that the quantity Δ_t defined by Eq. (6.51), being positive from its physical interpretation, turns out to be Gaussian (since it is expressed by the increment of the Wiener process); thus, Δ_t can also take negative values with positive probability. This is a consequence of randomization by a Gaussian white noise. However, as is known in practice for such a distribution, the *three sigma* rule is usually adopted. This means that *practically* possible values of Δ_t, for each fixed t, are in the interval $(m_{\Delta_t} - 3\sigma_{\Delta_t}, m_{\Delta_t} + 3\sigma_{\Delta_t})$; the probability that $|\Delta_t - m_{\Delta_t}| > 3\sigma_{\Delta_t}$ is less than 0.01.

31.3 Life Distribution

Since our governing stochastic differential equation (6.47) has, for $\rho > 1$, the coefficients that grow more quickly than linearly, the process $A(t)$ is defined on the random time interval $[t_0, \tau(\gamma)]$, where $\tau(\gamma)$ is the random time of *explosion* of the solution to infinity; $\tau(\gamma)$ can be regarded as the time at which unstable crack growth occurs, resulting in the ultimate fatigue failure of the specimen. Since our physical process is

$$A_t = \frac{1}{Y_t^{1/(\rho-1)}}, \tag{6.54}$$

where Y_t is given by Eq. (6.50), the *explosion* time $\tau(\gamma)$ is defined as

$$\begin{aligned}\tau(\gamma) = \inf\{t\colon &(\rho-1)c_1(t-t_0) \\ &+ (\rho-1)c_2[W(t) - W(t_0)] \geq \frac{1}{a_0^{\rho-1}}\}.\end{aligned} \tag{6.55}$$

The preceding *first passage time* is given by $\tau(\gamma) = t_0 + \tau_1(\gamma)$, where

$$\tau_1(\gamma) = \inf\{t\colon (\rho-1)c_1 t + (\rho-1)c_2 W_1(t) \geq \frac{1}{a_0^{\rho-1}}\}, \tag{6.56}$$

and $W_1(t)$ is a standard Wiener process, i.e., $P\{W_1(0) = 0\} = 1$ and its variance is equal to one.

Introducing the following denotations,

$$\begin{aligned}\alpha &= \frac{1}{a_0^{\rho-1}(\rho-1)c_2} > 0, \\ \beta &= \frac{c_1}{c_2} = \frac{m_x}{\sqrt{2D}} > 0,\end{aligned} \tag{6.57}$$

we have

$$\tau_1(\gamma) = \inf\{t\colon W(t) \geq \alpha - \beta t\}. \tag{6.58}$$

Hence, $\tau_1(\gamma)$ characterizes a time of a first passage of the standard Wiener process through the line: $\alpha - \beta t$. It has been shown in [180] and [188] that the probability density of the random variable $\tau_1(\gamma)$ is

$$f_{\tau_1}(t) = \frac{\alpha}{\sqrt{2\pi}\, t^{3/2}} \exp\left[-\frac{1}{2}\frac{(\alpha-\beta t)^2}{t}\right], \tag{6.59}$$

which is the inverse Gaussian distribution. (See Eq. (2.36).) This distribution has previously been hypothesized in reliability theory as a possible lifetime model. (See Chhikara and Folks [59].) Here, it has been derived from the stochastic differential equation model for fatigue crack growth. In the analysis of random fatigue crack growth, the inverse Gaussian distribution has also been derived (as a life distribution) by Ditlevsen [162] by an alternate approach based on an incremental version of the Paris–Erdogan equation.

The estimation of parameters of the model presented in the preceding can be carried out by use of known methods of statistics. Since we have explicit expressions for the probability distributions of the crack size and life of the specimen, the parameters can be estimated, for example, by the maximum likelihood method. (See Newby [181].)

However, in the situation when a convenient analytical expression for the probability distribution of crack size is not available (e.g., when boundary conditions are explicitly considered), estimation of model parameters is more complex and has been considered in [176] and [191].

31.4 Extensions

In the analysis just presented, we showed the simplest case that leads to Markovian description of fatigue crack growth; but, as we already underlined, the Gaussian white noise as a randomizing process has essential deficiencies. Since dA/dt is a nonnegative physical quantity, the process $X(t, \gamma)$ should have the property: $X(t, \gamma) > 0$ for all t and γ. This dictates, however, that it cannot be a Gaussian process; the probability distribution of $X(t, \gamma)$ should be concentrated on the positive half-line.

The correlation between the values of the process $X(t, \gamma)$ at different time instants is also important to consider. If $X(t, \gamma)$ is uncorrelated at each instant of time (i.e., a Gaussian white noise), then the smallest statistical dispersion of crack size is obtained, whereas if $X(t, \gamma)$ is considered to be perfectly correlated (i.e., a random variable), then the largest statistical dispersion is obtained. Experimental evidence indicates that the correct situation is somewhere in between. In this situation, an important problem arises (first stated and discussed in the paper [127]): How can we relax the Gaussian white noise assumption and include necessary correlation information without losing the advantages of Markov process theory in the analysis of random fatigue crack growth?

As indicated in [127], there are two approaches toward more realistic randomization of empirical fatigue crack growth within Markov diffusion processes theory. The first possibility lies, generally speaking, in application of asymptotic analysis to the randomized fatigue crack equation. This approach allows us to introduce correlation of the values of the randomizing process $X(t, \gamma)$. The initial question (discussed in the paper [127] and, independently, by Lin and Yang [176], [177], where specific analysis was provided) is: When can the process $A(t)$ governed by the general stochastic equation of the form,

$$\frac{dA}{dt} = G[A(t), X(t, \gamma)], \tag{6.60}$$

be considered as a diffusion Markov process?

Physically, $A(t)$ can be regarded as approximately Markovian for time intervals greater than Δ, where Δ is such that $\tau_k \leq \Delta$, and τ_k is the correlation time of process $X(t, \gamma)$.

The relation $\tau_k \leq \Delta$ means that increments of $A(t)$ on non-overlapping long time intervals Δ shall be practically independent. More detailed analysis indicates also that the motion of the crack tip should be sufficiently *inertial*, i.e., $A(t)$ and dA/dt should vary slowly within intervals where correlation is significant; in such a case, the increment of $A(t)$ within small time intervals will be small — which is a characteristic feature of diffusion Markov processes.

If the conditions so described are satisfied, the process $A(t)$ governed by Eq. (6.60), in particular by Eq. (6.35), can be approximated by the diffusion Markov process that is governed by the Itô stochastic differential equation (6.40) with the following drift and diffusion coefficients:

$$\begin{aligned} m^*(A) &= \langle G(A, X)\rangle + \int K_{G'G_\tau}(A, \tau)\, d\tau, \\ \sigma^2(A) &= 2\int \langle \underset{\sim}{G}(A, X(\tau))\, G(A, X(t+\tau))\rangle d\tau, \end{aligned} \tag{6.61}$$

where $\langle G(A, X)\rangle$ is a mean value of the right-hand side of Eq. (6.60) when A is fixed; $\underset{\sim}{G}(A, X(t))$ is a fluctuation of $G(A, X(t))$, i.e.,

$$\underset{\sim}{G}(A, X(t)) = G(A, X(t)) - \langle G(A, X(t))\rangle, \tag{6.62}$$

and $K_{G'G_\tau}(A, \tau)$ is a mutual correlation function of the derivative of the fluctuation part of G (with respect to A) and G in different time instants. More detailed formulae can be found in [127], [176], and [177].

The preceding heuristic reasoning implies that the effect of the past correlation of the process $X(t, \gamma)$ is lumped (approximately) and placed at the present. This idea, when rigorously formulated, leads to the averaging method for stochastic differential equations developed by Stratonovich and Khasminski. (For a systematic presentation, see [77].)

Randomization of the Paris–Erdogan equation along the lines presented in the preceding, i.e.,

$$\begin{aligned} \frac{dA}{dt} &= C_0 A^{\rho} X(t, \gamma) \\ &= C_0 A^{\rho} [m_X + X_1(t, \gamma)] \\ &= c_1 A^{\rho} + C_0 A^{\rho} X_1(t, \gamma), \end{aligned} \tag{6.63}$$

leads to the Itô differential equation whose coefficients coincide with Eq. (6.45) when constant c_2 is replaced by the correlation time of the process $X(t, \gamma)$, defined as

$$l = \int_0^{\infty} K_{X_1}(\tau)\, d\tau. \tag{6.64}$$

As already mentioned, a detailed analysis of the random fatigue crack growth process modelled by Eq. (6.63) was performed by Lin and Yang (see [176] and [177]), where a comparison of the theoretical probability distributions with their empirical counterparts was shown. More recently, Solomos [189] reported the analysis of the first-passage time distribution for the Itô stochastic differential equation model with reflecting and absorbing boundary conditions at the ends of interval the $[a_0, a_{cr}]$.

Although the use of the concept of averaging described before gives some possibility to account for the correlation of real processes affecting crack growth, it still includes the Gaussian distribution assumption (because of the increments of the Wiener process in the Itô stochastic equation), which in general results in the possibility of negative value of $dA(t)/dt$.

To account for the mutual correlation of the values of the randomizing process and to exclude the possibility (even of small probability) of negative values of the crack growth rate $dA(t)/dt$, we can construct a system of stochastic differential equations in which, in addition to $A(t)$, the randomizing process occurs as an unknown; this is a second possibility mentioned at the beginning of this section and will constitute the content of the next section.

32. Two-State Crack Growth Model

Let us introduce an auxiliary random process $Z(t)$, which is a filtered white noise process, and a deterministic function $g(z)$, which transforms the process $Z(t)$ into a positive process $X(t)$ (e.g., $g(z)$ could be taken as an exponential function). Then the random process $X(t)$ in Eqs. (6.35) and (6.63) can be represented by $X(t) = g(Z(t))$, and a coupled system of stochastic differential equations is obtained,

$$\begin{aligned} \frac{dA}{dt} &= F[A(t)]\, g[Z(t)] \\ \frac{dZ}{dt} &= Q[Z(t)] + \xi(t,\gamma) \\ A(0) = A_0(\gamma), &\quad Z(0) = Z_0(\gamma), \end{aligned} \tag{6.65}$$

where $\xi(t,\gamma)$ is a Gaussian white noise process with intensity πS_0, and $A_0(\gamma)$, $Z_0(\gamma)$ are general random variables characterizing the initial crack size and the initial state of the auxiliary random process, respectively. Because both F and g are defined as positive functions of their respective arguments, the possibility of negative crack growth rates encountered in the white noise models therefore has been eliminated. Q can then be appropriately constructed such that the random fluctuations in crack growth rate have the proper correlation structure.

If the system of equations in Eq. (6.65) are interpreted in the Itô sense, the vector process $[A(t), Z(t)]^T$ is a diffusion Markov process, and all of the attendant theory is available for the description of the random crack size $A(t)$. Such an approach has been proposed and systematically elaborated on by Spencer and his co-workers. (See [190], [191], and [194].)

32.1 Reliability Problem Formulation

As discussed in Section 10.4, the transition probability density function $p = p(a, z, t | a_0, z_0, t_0)$ is the joint probability density function for the random crack size $A(t)$ and the auxiliary random process $Z(t)$, conditional on their initial values, and is governed by the Fokker–Planck–Kolmogorov equation,

$$\frac{\partial p}{\partial t} = -\frac{\partial}{\partial a}[F(a)\, g(z)\, p] - \frac{\partial}{\partial z}[Q(z)\, p] + \pi S_0 \frac{\partial^2 p}{\partial z^2}. \tag{6.66}$$

The transition density is also governed by the formal adjoint of Eq. (6.66), the backward Kolmogorov equation, given by

$$\frac{\partial p}{\partial t} = F(a_0)\, g(z_0) \frac{\partial p}{\partial a_0} + Q(z_0) \frac{\partial p}{\partial z_0} + \pi S_0 \frac{\partial^2 p}{\partial z_0^2}. \tag{6.67}$$

The cumulative distribution of times to reach a given crack size a_c, or conversely, the probability that a crack will not reach a critical crack size in a given time, is of great interest in the design of fatigue critical structures. Denoting the probability that the crack size is less than the critical crack size (conditional on the initial crack size and the initial value of $Z(t)$) as

$$\begin{aligned} R = R(t|a_0, z_0) &= P\{A(t) < a_c; t|a_0, z_0, t_0\} \\ &= \int_{a_0}^{a_c} \int_{-\infty}^{\infty} p(a, z, t|a_0, z_0, t_0)\, dz da, \end{aligned} \tag{6.68}$$

we see that $R(t|a_0, z_0)$ also satisfies the backward Kolmogorov equation (6.67), and we thus can write

$$\frac{\partial R}{\partial t} = F(a_0)\, g(z_0) \frac{\partial R}{\partial a_0} + Q(z_0) \frac{\partial R}{\partial z_0} + \pi S_0 \frac{\partial^2 R}{\partial z_0^2}. \tag{6.69}$$

One can notice that, due to the lack of a second derivative with respect to a_0 in Eq. (6.69), only one boundary condition needs to be specified in the a_0 direction in order that the initial-boundary value problem be well-posed. (See Fichera [170].) The initial and boundary conditions can be stated on physical grounds as follows:

$$R(0|a_0, z_0) = 1, \qquad 0 \le a_0 < a_c, \tag{6.70}$$

$$R(t|a_c, z_0) = 0, \qquad \forall\ z_0, \tag{6.71}$$

$$\lim_{z_0 \to \infty} R(t|a_0, z_0) = 0, \qquad 0 \le a_0 < a_c, \tag{6.72}$$

$$\lim_{z_0 \to -\infty} \frac{\partial R(t|a_0, z_0)}{\partial z_0} = 0, \qquad 0 \le a_0 < a_c. \tag{6.73}$$

Further discussion of the initial and boundary conditions follows.

The initial condition given in Eq. (6.70) is rather evident; the reliability of the structural component is assumed to be equal to one before being put into service (i.e., the initial crack size is considered to be below the critical crack size at the time the structural member is put into service).

The boundary condition given by Eq. (6.71) is also easy to envision and simply indicates that the reliability of the component is zero for all z_0 when the initial crack size is the critical crack size (i.e., $a_0 = a_c$).

We assume that the function $Q(z)$ is such that the domain of $Z(t)$ is the interval $(-\infty, \infty)$, that $g(z)$ is a one-to-one mapping onto the interval $(0, \infty)$, and that $\partial g/\partial z \geq 0$. Therefore, because $F(\cdot)$ is positive for all physically realizable arguments, the boundary condition specified in Eq. (6.72) can be justified by examining Eq. (6.65) and noting that as $z_0 \to \infty$, so also does $dA/dt \to \infty$. Since the random process $Z(t)$ is continuous, $Z(t)$ and thus dA/dt cannot instantaneously make large changes. Thus, as $z_0 \to \infty$, $A(t)$ will grow very rapidly, and in the limit, the reliability will go to zero (i.e., $R(t|a_0, z_0) \to 0$).

The third boundary condition is more difficult to envision. Because of the assumptions on $g(\cdot)$ stated in the previous paragraph, as $z_0 \to -\infty$, then $dA/dt \to 0$. Thus, there is no crack growth. As a result, one might expect the reliability to go to one (i.e., the time to reach the critical crack size goes to infinity) as the initial value of $Z(t)$ goes to negative infinity. However, there is a limiting factor. Once the time to reach the critical crack size a_c becomes much larger than the correlation time for $Z(t)$, the initial value of $Z(t)$ will have little effect on the reliability; that is, the increment of the reliability approaches zero as the initial value for $Z(t)$ goes to negative infinity, implying that the derivative of the reliability goes to zero. This boundary condition has also been demonstrated through Monte Carlo simulation [190].

The initial-boundary value problem given by Eqs. (6.69)–(6.73) is therefore defined over a semi-infinite two-dimensional domain. Here, we have constructed the problem based upon a two-dimensional Markov process such that negative crack growth rates are impossible and the correlation structure of the random crack growth rate can be included without employing stochastic averaging.

In this problem formulation, the solution is a function of the initial value of the auxiliary random process, $Z(t)$. To eliminate the dependence on z_0, one can assume a stationary start condition for $Z(t)$ and make use of the theory of total probability to obtain

$$R(t|a_0) = \int_{-\infty}^{\infty} R(t|a_0, z_0) f_Z(z_0)\, dz_0, \tag{6.74}$$

where $f_Z(z_0)$ is the stationary probability density function for $Z(t)$. In this final step, we have completed the formulation to obtain the reliability of a component against fatigue crack growth conditional on the initial flaw size.

Remark: One of the most attractive features of this approach is that the solution is obtained as a function of the initial crack size. This allows one to readily assess the effect of initial fatigue quality on the reliability calculations through the use of initial crack size distributions. For example, if the distribution of initial crack sizes is given by $f_{A_0}(a_0)$, then we can determine the unconditional reliability as a function of time via

$$R(t) = \int_0^{a_c} R(t|a_0) f_{A_0}(a_0)\, da_0, \tag{6.75}$$

and thus the effect of initial flaw size distribution is readily assessed.

32.2 Statistical Moments of Time to Reach a Critical Crack Size

One of the problems that is of interest to many researchers is determination of the statistical moments of a random process, not only because they are more readily obtained than the complete probability distribution function, but because they generally give the main features of a random process. In this section, we formulate the boundary value problem for the statistical moments of the time to reach a critical crack size.

Denoting the ordinary moment of the random time to reach a critical crack size a_c as

$$T^n(a_0, z_0) = \int_0^{\infty} t^n \frac{\partial R(t|a_0, z_0)}{\partial t} dt, \tag{6.76}$$

we can multiply Eq. (6.69) by t^{n-1} and integrate with respect to time to obtain a recursive set of boundary value problems given by

$$-nT^{n-1} = F(a_0)\, g(a_0) \frac{\partial T^n}{\partial a_0} + Q(z_0) \frac{\partial T^n}{\partial z_0} + \pi S_0 \frac{\partial^2 T^n}{\partial z_0^2}, \quad n = 1,2, \dots, \tag{6.77}$$

where, by definition, $T^0(a_0, z_0) = 1$, and the boundary conditions are given respectively by

$$T^n(a_c, z_0) = 0, \qquad \forall\, z_0, \tag{6.78}$$

$$\lim_{z_0 \to \infty} T^n(a_0, z_0) = 0, \qquad 0 \le a_0 < a_c, \tag{6.79}$$

$$\lim_{z_0 \to -\infty} \frac{\partial T^n(a_0, z_0)}{\partial z_0} = 0, \qquad 0 \le a_0 < a_c. \tag{6.80}$$

The boundary conditions were obtained directly from Eq. (6.76) and the boundary conditions for the reliability problem.

Oftentimes, we are more interested in the calculation of the central moments than the ordinary moments. The second central moment (i.e., the variance V) is given in terms of the ordinary moments as

$$V = T^2 - (T^1)^2. \tag{6.81}$$

The boundary value problem can then be formulated directly in terms of the second central moments as

$$-2\pi S_0 \left(\frac{\partial T^1}{\partial z_0}\right)^2 = F(a_0)\, g(a_0) \frac{\partial V}{\partial a_0} + Q(z_0) \frac{\partial V}{\partial z_0} + \pi S_0 \frac{\partial^2 V}{\partial z_0^2}, \tag{6.82}$$

where the boundary conditions are

$$V(a_c, z_0) = 0, \qquad \forall\, z_0, \tag{6.83}$$

$$\lim_{z_0 \to \infty} V(a_0, z_0) = 0, \qquad 0 \le a_0 < a_c, \tag{6.84}$$

$$\lim_{z_0 \to -\infty} \frac{\partial V(a_0, z_0)}{\partial z_0} = 0, \qquad 0 \le a_0 < a_c. \tag{6.85}$$

The boundary conditions for the equations governing the central moments are also obtained directly from Eq. (6.76) and the boundary conditions for the reliability problem. The preceding manipulations could be carried out for the higher-order central moments in a similar manner.

As stated in the previous section, the dependence of the solution on z_0 can be eliminated by assuming a stationary start condition for the random process $Z(t)$ and making use of the theorem of total probability. The solutions, however, are also a function of the initial crack size. Thus, either the actual initial crack size or the initial crack size distribution can be used to obtain the unconditional moments of the random time to reach a critical crack size.

Although analytical solution of the boundary value problems presented in the preceding is unavailable at this time, a simple and efficient numerical solution has been reported by Enneking et al. [168] and will be presented in the next section.

32.3 Numerical Solution

The solution procedure is a hybrid finite element–finite difference scheme. The solution process begins at the known zero boundary (i.e., when $a_0 = a_c$) at time $t = 0$ and uses a variable-weighted finite difference marching technique to move forward in time and backward in initial crack size. The solution in the z_0 direction, for each instant in time and row in a_0, is determined using a finite element approach. The infinite boundary conditions in the z_0 direction are placed at a distance sufficiently removed from the domain of interest so that the solution is not affected. This approach is graphically illustrated in Fig. 6.2.

The governing partial differential equation for the reliability problem (6.69) can be discretized in time using a variable-weighted finite difference approach with weighting parameter θ as

$$\frac{R^{t+\Delta t} - R^t}{\Delta t} = F(a_0)\, g(z_0) \frac{\partial R^{t+\theta}}{\partial a_0} + Q(z_0) \frac{\partial R^{t+\theta}}{\partial z_0} + \pi S_0 \frac{\partial^2 R^{t+\theta}}{\partial z_0^2}, \quad (6.86)$$

where R^t is the solution at time t. Thus, by using a variable-weighted finite difference scheme, the terms in Eq. (6.86) are approximated as some weighted average over the interval Δt, as opposed to being approximated

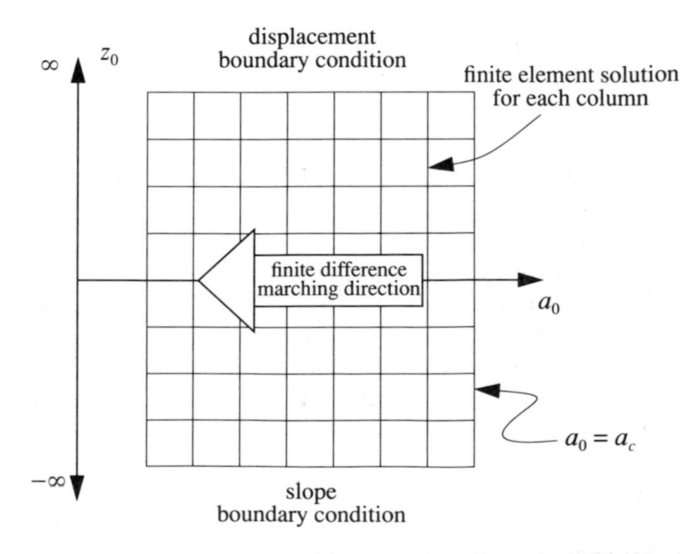

Figure 6.2. Schematic of problem solution methodology for each time step.

entirely at t or $t+\Delta t$. The choice of $\theta = 0.5$ yields the well-known Crank–Nicholson scheme. It can also be shown that this method will be stable for values of $\theta_1 \geq 0.5$. Stability when $\theta < 0.5$ is dependent on the value chosen for Δt. A value of $\theta = 0.5$ is selected to minimize the truncation error in the finite difference approximation.

Defining $R^{t+\theta} = \theta R^{t+\Delta t} + (1-\theta) R^t$, Eq. (6.86) can be written as

$$R^{t+\Delta t} - R^t = \Delta t\ F(a_0)\, g(z_0) \left(\theta \frac{\partial R^{t+\Delta t}}{\partial a_0} + (1-\theta) \frac{\partial R^t}{\partial a_0}\right)$$
$$+ \Delta t\ Q(z_0) \left(\theta \frac{\partial R^{t+\Delta t}}{\partial z_0} + (1-\theta) \frac{\partial R^t}{\partial z_0}\right) \tag{6.87}$$
$$+ \Delta t\ \pi S_0 \left(\theta \frac{\partial^2 R^{t+\Delta t}}{\partial z_0^2} + \theta \frac{\partial^2 R^t}{\partial z_0^2}\right).$$

The problem can be simplified by letting $\delta R = R^{t+\Delta t} - R^t$ to obtain

$$\delta R - \theta \Delta t \left(F(a_0)\, g(z_0) \frac{\partial \delta R}{\partial a_0} + Q(z_0) \frac{\partial \delta R}{\partial z_0} + \pi S_0 \frac{\partial^2}{\partial z_0} \delta R\right) = G^t, \tag{6.88}$$

where G^t is a function of the solution at the previous time step and is given by

$$G^t = \Delta t \left(F(a_0)\, g(z_0) \frac{\partial R^t}{\partial a_0} + Q(z_0) \frac{\partial R^t}{\partial z_0} + \pi S_0 \frac{\partial^2 R^t}{\partial z_0^2}\right). \tag{6.89}$$

Equation (6.88) can be rewritten in the general form

$$c_1 \frac{\partial^2 f}{\partial z_0^2} + c_2 \frac{\partial f}{\partial z_0} + c_3 f + c_4 = c_5 \frac{\partial f}{\partial a_0}, \tag{6.90}$$

where $f = \delta R$, $c_1 = -\theta \Delta t \pi S_0$, $c_2 = -\theta \Delta t Q(z_0)$, $c_3 = 1$, $c_4 = -G^t$, and $c_5 = \theta \Delta t F(a_0)\, g(z_0)$. We again discretize Eq. (6.90) with respect to a_0 using the same variable-weighted scheme as employed for discretization in time to obtain

$$\frac{f^{a+\Delta a} - f^a}{\Delta a} - \frac{c_1}{c_5} \frac{\partial^2 f^{a+\theta}}{\partial z_0^2} - \frac{c_2}{c_5} \frac{\partial f^{a+\theta}}{\partial z_0} - \frac{c_3}{c_5} f^{a+\theta} = \frac{c_4}{c_5}. \tag{6.91}$$

Then, expressing $f^{a+\theta}$ as a weighted average $\theta f^{a+\Delta a}+(1-\theta)f^a$ and defining $\delta f = f^{a+\Delta a} - f^a$, we can write

$$\frac{\partial^2}{\partial z_0^2}\delta f + \frac{c_2}{c_1}\frac{\partial \delta f}{\partial z_0} + \frac{c_3 \Delta a - c_5/\theta}{\Delta a \; c_1}\delta f = -\frac{1}{\theta}\left(\frac{\partial^2 f^a}{\partial z_0^2} + \frac{c_2}{c_1}\frac{\partial f^a}{\partial z_0} + \frac{c_3}{c_1}f^a + \frac{c_4}{c_1}\right). \tag{6.92}$$

Notice that this equation is similar in form to a standard one-dimensional convection diffusion equation with the addition of a linear term and a nonzero right-hand side that is dependent only on the solution for the row at the previous increment in a_0. To obtain a solution to Eq. (6.92), we cast it into the weak form using arbitrary weighting functions W_i, defined to be zero on the boundaries. In this simple one-dimensional finite element case, integration by parts is used to eliminate the second derivative terms so that linear interpolation functions N_n can be readily applied. Linear functions are also used for the weighting functions. Thus, the weak form can be expressed as follows:

$$\begin{aligned}&\int_{-\infty}^{\infty} W_i\left(\frac{\partial^2}{\partial z_0^2}\delta f + \frac{c_2}{c_1}\frac{\partial \delta f}{\partial z_0} + \frac{c_3 \Delta a - c_5/\theta}{\Delta a \; c_1}\delta f\right)dz_0 \\ &\qquad = \frac{-1}{\theta}\int_{-\infty}^{\infty} W_i\left(\frac{\partial^2 f^a}{\partial z_0^2} + \frac{c_2}{c_1}\frac{\partial f^a}{\partial z_0} + \frac{c_3}{c_1}f^a + \frac{c_4}{c_1}\right)dz_0,\end{aligned} \tag{6.93}$$

or, after partial integration,

$$\begin{aligned}&-\int_{-\infty}^{\infty}\left[\frac{\partial W_i}{\partial z_0}\frac{\partial \delta f}{\partial z_0} - \frac{W_i c_2}{c_1}\frac{\partial \delta f}{\partial z_0} - \frac{W_i\,(c_3 \Delta a - c_5/\theta)\,\delta f}{\Delta a \; c_1}\right]dz_0 + W_i\frac{\partial \delta f}{\partial z_0}\bigg|_{-\infty}^{\infty} \\ &= \frac{-1}{\theta}\left[\int_{-\infty}^{\infty}\left\{\left(\frac{W_i c_2}{c_1} - \frac{\partial W_i}{\partial z_0}\right)\frac{\partial f^a}{\partial z_0} + \frac{W_i c_3}{c_1}f^a + \frac{W_i c_4}{c_1}\right\}dz_0 + W_i\frac{\partial f^a}{\partial z_0}\bigg|_{-\infty}^{\infty}\right].\end{aligned} \tag{6.94}$$

Note that the boundary terms in Eq. (6.94) are equal to zero due to the zero boundary conditions and the generally accepted practice of selecting weighting functions that are zero on the boundaries.

Discretizing in space by letting $f = \Sigma N_n(z_0) f_n$, Eq. (6.94) becomes

$$\sum\left\{\int_{-\infty}^{\infty}\left[\frac{\partial W_i}{\partial z_0}\frac{\partial N_n}{\partial z_0} - \frac{W_i c_2}{c_1}\frac{\partial N_n}{\partial z_0} - \frac{W_i(c_3\Delta a - c_5/\theta)N_n}{\Delta a\ c_1}\right]dz_0\right\}\delta f_n$$

$$= \sum\frac{1}{\theta}\left\{\int_{-\infty}^{\infty}\frac{W_i c_4}{c_1}dz_0 + \left[\int_{-\infty}^{\infty}\left\{\left(\frac{W_i c_2}{c_1} - \frac{\partial W_i}{\partial z_0}\right)\frac{\partial N_n}{\partial z_0} + \frac{W_i c_3}{c_1}N_n\right\}dz_0\right]f_n^a\right\}. \tag{6.95}$$

After transforming to the isoparametric coordinates, η, Eq. (6.95) can be expressed in matrix form as

$$[\mathbf{K}_1 - \mathbf{K}_2]\,\delta\mathbf{f} = \mathbf{P} + \frac{1}{\theta}[-\mathbf{K}_1 + \mathbf{H}]\,\mathbf{f}^a, \tag{6.96}$$

where

$$\mathbf{K}_1 = \int_{-1}^{1}\left\{\mathbf{W}' - \frac{c_2}{c_1}\mathbf{W}\right\}^{\mathrm{T}}\mathbf{N}'|\mathbf{J}|\,d\eta, \tag{6.97a}$$

$$\mathbf{K}_2 = \int_{-1}^{1}\frac{c_3\Delta a - c_5/\theta}{\Delta a\ c_1}\mathbf{W}^{\mathrm{T}}\mathbf{N}|\mathbf{J}|\,d\eta, \tag{6.97b}$$

$$\mathbf{H} = \int_{-1}^{1}\frac{c_3}{c_1}\mathbf{W}^{\mathrm{T}}\mathbf{N}|\mathbf{J}|\,d\eta, \tag{6.97c}$$

$$\mathbf{P} = \int_{-1}^{1}\frac{c_4}{\theta c_1}\mathbf{W}^{\mathrm{T}}|\mathbf{J}|\,d\eta, \tag{6.97d}$$

and the superscript T denotes transpose. In Eqs. (6.97a)–(6.97d), **W** and **N** are column matrices that contain the weight and shape functions; **W**' and **N**' are matrices of the derivatives with respect to z_0 of the weighting and shape functions, respectively; and $|\mathbf{J}|$ is the determinant of the Jacobian matrix, half the element length in this one-dimensional case. Since the coefficients c_1 through c_5 may be functions of a_0 and z_0, they should be appropriately interpolated in the calculations.

Once δf is obtained, it can be back-substituted to obtain δR and, ultimately, the sought reliability function $R\,(t|\,a_0, z_0)$.

We also remark that the governing differential equations for the moments of time to reach a critical crack size (i.e., Eqs. (6.77) and (6.82)) are of the form of general equation (6.90). For example, we would have $c_1 = \pi S_0$, $c_2 = Q(z_0)$, $c_3 = 0$, $c_4 = nT^{n-1}$, $c_5 = -F(a_0)g(z_0)$. Therefore, the numerical scheme derived previously is also applicable to moment problems. (See [168].)

With solution of the two-state problem in hand, we turn our attention to the problem of parameter estimation. The traditional approach to parameter estimation [202] employs an incremental polynomial to estimate dA/dt from the experimental data and then performs a least-squares fit of the data to obtain the sought parameters. The deficiencies of this approach and alternative parameter estimation methods will be discussed in the next section.

32.4 Parameter Estimation

Use of any model requires appropriate procedures for estimating the parameters of the model. To effectively discuss parameter estimation procedures, we first focus our attention on estimating the parameters of the randomized Paris–Erdogan equation,

$$\frac{dA}{dt} = C(\Delta K)^m X(t), \tag{6.98}$$

where, as is customary in the literature, $X(t)$ is assumed to be a stationary lognormal random process. The parameters that need to be estimated in this model are then C, m, and the characteristics of $X(t)$. We assume that the data are available in the form of observed values of crack length a_i at a given time instant t_i.

We present two approaches to the problem of parameter estimation in this section. The first, termed the least-squares approach, is an extension of the traditional method for estimating the parameters for the deterministic Paris–Erdogan equation to its stochastic analog (6.98). (See [202].) This approach requires that estimates of dA/dt be obtained from the a versus t data. The second approach presented obtains the model parameters without requiring that estimates of dA/dt be determined from the data.

32.4.1 Least-Squares Estimation

Let us take the logarithm of both sides of Eq. (6.98) to obtain

$$\ln\left(\frac{dA}{dt}\right) = m\ln\Delta K + \ln C + Z(t), \tag{6.99}$$

where $Z(t) = \ln X(t)$ is a stationary Gaussian process. For each sample point (a_i, t_i) in the experimental data, a random residual can be defined as

$$z_i = \ln \left(\frac{da}{dt}\right)_i - m \ln \Delta K_i - \ln C. \tag{6.100}$$

One criterion for choosing m and C is to minimize the sum of the squares of Z_i. This naturally leads to a least-square regression analysis to estimate m and $\ln C$. Defining the quantities $\underset{\sim}{x} = \ln \Delta K$ and $\underset{\sim}{y} = \ln (dA/dt)$, the parameter estimates are

$$m = \frac{\Sigma \underset{\sim}{x}_i \underset{\sim}{y}_i - n\bar{\underset{\sim}{x}}\ \bar{\underset{\sim}{y}}}{\Sigma \underset{\sim}{x}_i^2 - n\bar{\underset{\sim}{x}}} \tag{6.101}$$

$$\ln C = \bar{\underset{\sim}{y}} - m\bar{\underset{\sim}{x}},$$

where the subscript i refers to a particular data point (a_i, t_i), $\Sigma = \Sigma_{i=1}^{n}$, and $\bar{\underset{\sim}{x}}, \bar{\underset{\sim}{y}}$ are the sample means of $\underset{\sim}{x}$ and $\underset{\sim}{y}$, respectively.

Because $Z(t)$ is assumed to be stationary, σ_Z^2 is constant and can be estimated by taking the sample variance of the difference between the actual data and the regression line. The stationary variance of $Z(t)$ is then given by

$$\sigma_Z^2 = \frac{1}{n-2} \sum_{i=1}^{n} z_i^2. \tag{6.102}$$

The factor of $1/(n-2)$ is required to give an unbiased estimate of the sample variance of $Z(t)$ obtained in the regression analysis.

We wish to point out that there is an inherent bias introduced by the log–log regression, or least-squares, approach discussed in this section. As pointed out by Ferguson [169], the arithmetic mean, instead of the geometric mean, is obtained for dA/dt using this method.

To use the least-squares method to estimate the parameters, calculation of the derivative of the crack length with respect to time is required. Several approaches can be employed to determine $(da/dt)_i$ for use in the preceding analysis. For deterministic analyses, the American Society for Testing and Materials has advocated two techniques for these calculations: the secant method and the incremental polynomial method. (See [158].) These two methods will be elaborated upon in the next few paragraphs.

The secant method approximates dA/dt by passing a line through the points (a_i, t_i) and (a_{i+1}, t_{i+1}) and using the slope of the secant line as the crack growth rate at $(a_i + a_{i+1})/2$, i.e.,

$$\left.\frac{da}{dt}\right|_{\bar{a}} \approx \frac{a_{i+1} - a_i}{t_{i+1} - t_i} , \tag{6.103}$$

where $\bar{a} = (a_{i+1} + a_i)/2$. (See Fig. 6.3.)

There are difficulties in using the secant method to estimate da/dt. Using the secant estimates in Eq. (6.101) to determine the parameters in a stochastic model will generally result in a model that overpredicts the structural fatigue life. This effect is illustrated in Fig. 6.3, where it can be seen that the slope of the secant line will generally be smaller than the slope of the true tangent line at $\bar{a}$. This implies that the secant estimates of (da/dt) will underestimate the true crack growth rates, thus accounting for the overprediction of specimen fatigue life.

The other technique recommended by ASTM for estimating da/dt from the a versus t data is the incremental polynomial method. In this approach, a second-order polynomial is fit to three, five, seven, or nine successive data points using least-squares. The rate of crack growth is then obtained from the derivative of the polynomial at the middle data point. The more data points employed, the more *smoothed* the estimates of the (da/dt) become, i.e., the dispersion in the data is artificially reduced by fitting the polynomial to the data. As a result, employing the incremental polynomial method to estimate the parameters in a stochastic model can produce poor results [191].

As indicated previously, methods that differentiate the a versus t data to obtain the crack growth rate da/dt can introduce considerable error, depending on the particular numerical differentiation scheme utilized and the measurement increment, Δa. In fact, regression analysis of the differentiat-

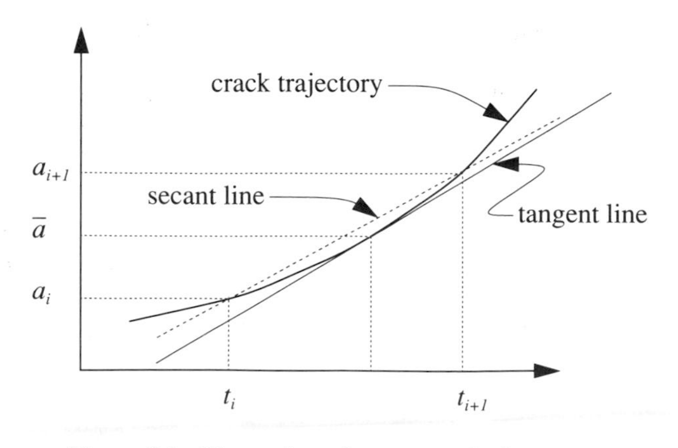

Figure 6.3. Illustration of secant method.

ed data followed by subsequent integration of the resulting crack growth rate equations leads to inconsistent reproduction of the actual data [184]. The next section will present an approach to alleviate some of these difficulties.

32.4.2 Modified Finite Integral Approach

Ostergaard and Hillberry [184] developed a finite integral method to avoid the numerical differentiation of the original experimental data in the estimation of parameters in the deterministic fatigue crack growth law. In this approach, the deterministic analog of Eq. (6.98) is separated and integrated to obtain

$$\int_{a_0}^{a_i} \frac{da}{C(\Delta K)^m} = \int_0^{T_{ri}} dt = T_{ri}, \tag{6.104}$$

where a_0 is the initial crack size, and T_{ri} is the time predicted by the model at which the crack will have a length a_i.

A sum of the squared errors $\bar{e}$ can then be defined as

$$\bar{e} = \sum_{i=1}^{N_{\Delta a}} (T_{ai} - T_{ri})^2, \tag{6.105}$$

where T_{ai} is the cumulative cycle count from the crack growth data, and $N_{\Delta a}$ is the number of crack intervals into which the data are broken. The optimal set of model parameters are defined as those that minimize the squared error $\bar{e}$.

This approach can be modified for parameter estimation in stochastic fatigue crack growth models [191]. Here, Eq. (6.98) is separated, integrated, and averaged over the ensemble to obtain

$$\int_{a_0}^{a_i} \frac{da}{C(\Delta K)^m} = \left\langle \int_0^{T_{ri}} X(t)\, dt \right\rangle, \tag{6.106}$$

where a_i is fixed, thus making the left-hand side of the equation a constant, and $\langle \cdot \rangle$ denotes expectation with respect to the ensemble. Employing the mean value theorem, the integral on the right-hand side of Eq. (6.106) can be rewritten as

$$\int_0^{T_{ri}} X(t)\, dt = X(s)\, T_{ri}, \tag{6.107}$$

where s is on the interval $[0, T_{ri}]$. Thus,

$$\left\langle \int_0^{T_{ri}} X(t)\, dt \right\rangle = \langle X(s)\, T_{ri} \rangle = \langle X(s) \rangle \langle T_{ri} \rangle + \rho \sigma_X \sigma_T, \tag{6.108}$$

where ρ is the correlation between $X(s)$ and T_{ri} (n.b., $|\rho| \leq 1$), σ_X is the standard deviation of the stationary process $X(t)$, and σ_T is the standard deviation of the random variable T_{ri}.

One can argue that the term $\rho \sigma_X \sigma_T$ is negligible as compared to $\langle X(s) \rangle \langle T_{ri} \rangle$ by noting that $\rho \to 0$ at T_{ri} increases and that practical experience indicates that $\sigma_X < 1$ and $\sigma_T \ll \langle T_{ri} \rangle$. Therefore, Eq. (6.106) can be well-approximated as

$$\int_{a_0}^{a_i} \frac{da}{C(\Delta K)^m} = \langle X(t) \rangle \langle T_{ri} \rangle. \tag{6.109}$$

This approximation has been verified with several extensive data sets in [191] and [194].

The sum of the squared errors for the stochastic parameter estimation problem can then be defined as

$$e = \sum_{i=1}^{N_{\Delta a}} (\langle T_{ai} \rangle - \langle T_{ri} \rangle)^2, \tag{6.110}$$

where $\langle T_{ai} \rangle$ is the mean cumulative cycle count obtained from the data, and $\langle T_{ri} \rangle$ is calculated from Eq. (6.109).

The optimum vales of the parameters in the stochastic model are obtained by minimizing the sum of the squared errors $\bar{e}$. This can be effectively done by using a numerical search algorithm, such as the IMSL routing ZXSSQ [172].

In the case of the randomized Paris–Erdogan model (6.98), the parameters C, m, and $\langle X(s) \rangle$ are determined from the modified finite integral method. However, there are two more parameters to be determined in the model, σ_X and a correlation parameter for the random process $X(t)$. While numerical differentiation of the fatigue crack growth data is not required in the modified finite integral method to determine the material constants, it is required to obtain an estimate of the variance σ_X and can be calculated as follows. We can write $X(t)$ as

$$X(t) = \frac{dA/dt}{C(\Delta K)^m}, \tag{6.111}$$

and the variance can be expressed as

$$\sigma_X^2 = \frac{1}{n-2}\sum_{i=1}^{n}\left(\frac{(dA/dt)_i}{C(\Delta K_i)^m} - \langle X(t)\rangle\right). \tag{6.112}$$

The secant method will give the largest estimation of σ_X^2 among those methods recommended by ASTM [158]; therefore, to be conservative, the secant method is recommended for calculation of $(dA/dt)_i$. Determination of the correlation parameter is dependent upon the particular form of the function $Q(Z(t))$ in Eq. (6.65) and, therefore, will be discussed with regard to particular examples.

The next section confronts the two-state Markov process model with experimental data to demonstrate the validity and accuracy of the model.

32.5 Correlation with Experimental Results

Data are essential for validation of any mathematical model. The fatigue crack growth data set reported by Virkler [54] (see Section 7.1) is one of high replication under tightly controlled laboratory conditions and is well-suited for validation of probabilistic models. Replicate tests for 68 identical center cracked specimens were conducted and the a versus n are depicted in Fig. 1.11. To minimize statistical uncertainty in the tests, each of the specimens was taken from a single lot of 2024–T3 aluminum and each test was performed by the same operator on the same machine. The tests were run at a constant stress range of $\Delta S = 48.28$ MPa and with a stress ratio $R = 0.2$. This exacting data will be used to validate the stochastic model and to demonstrate its accuracy and versatility.

For the study of fatigue crack growth under constant-amplitude loading, the crack growth rate is commonly considered as a function of the elastic stress intensity range given for a finite-width plate by

$$\Delta K = \Delta S\sqrt{\frac{\pi a}{\cos(\pi a/b)}}, \qquad \frac{a}{b} < 0.7, \tag{6.113}$$

with no more than 0.3% error, where ΔS is the stress range, and b is the width of the plate. A plot of $\ln(dA/dt)$ versus $\ln(\Delta K)$ for the Virkler data is depicted in Fig. 6.4.

While the predominantly linear trend in Fig. 6.4 indicates that the Paris–Erdogan-type model would well represent the data, Spencer et al. [191] has shown that the nonlinear trends are significant and can introduce systematic error into predicted results. To eliminate this systematic error, Kung and Ortiz [175] have suggested that a cubic polynomial fatigue crack growth law be used. In addition, the function $Q(\cdot)$ in Eq. (6.65) is chosen to be linear, and we assume that $X(t)$ is an exponentially correlated lognormal random process. The system of equations governing crack growth then becomes

$$\begin{aligned} \frac{dA}{dt} &= F[A(t)] \exp[Z(t)], \\ \frac{dZ}{dt} &= -\zeta Z(t) + \xi(t, \gamma), \end{aligned} \tag{6.114}$$

where ζ is the correlation parameter for the process $Z(t)$. For the cubic polynomial fatigue crack growth laws, $F[a(t)]$ is given by

$$F[a(t)] = \exp[\beta_0 + \beta_1(\ln\Delta K) + \beta_2(\ln\Delta K)^2 + \beta_3(\ln\Delta K)^3], \tag{6.115}$$

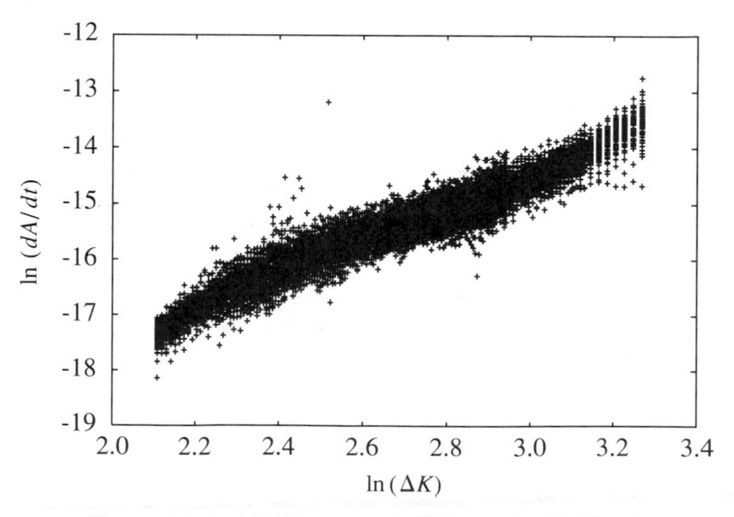

Figure 6.4. Ln(dA/dt) versus $\ln(\Delta K)$ for the Virkler data.

where $\Delta K = \Delta K(a)$ is given by Eq. (6.113). The parameters determined using the modified finite integral approach are (see [191]): $\beta_0 = -74.060$, $\beta_1 = 60.265$, $\beta_2 = -21.390$, $\beta_3 = 2.6348$, and $\langle X(t) \rangle = 1$.

The random residuals Z_i for this model are depicted in Fig. 6.5. We see that the residuals are zero mean and have a constant variance. Therefore, the multiplicative process $X(t) = \exp[Z(t)]$ is also stationary. The standard deviation of X is found via Eq. (6.112) to be $\sigma_X = 0.227$ and the correlation parameter is chosen to be $\zeta = 9\times10^{-5}$.

The boundary value problem in Eqs. (6.77)–(6.73) is solved for the statistical moments of time to reach a critical crack size a_c as a function of the initial crack size a_0. Figures 6.6–6.7 present the solutions for the mean and mean square, respectively, of the time to reach a critical crack size of 49.8 mm. The Virkler data is superimposed for comparison, and as can be seen, the predicted results are indiscernible from the data over the entire range of initial crack sizes.

Solving the boundary value problem in Eqs. (6.82)–(6.73) yields the variance of the time to reach a critical crack size a_c as a function of the initial crack size a_0. The square root of this result, the standard deviation, is depicted in Fig. 6.8. Again, the Virkler data is superimposed for comparison, and we see very good agreement with the results predicted by the model.

The boundary value problem for the reliability function given in Eqs. (6.69)–(6.73) is also solved to yield the distribution of the random time to reach a critical crack size. The results are given in Figs. 6.9–6.10. In Fig.

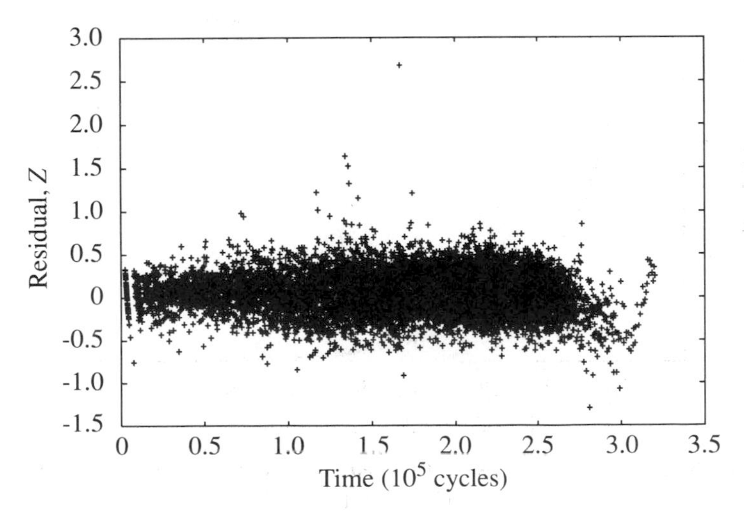

Figure 6.5. Residuals for the cubic polynomial model.

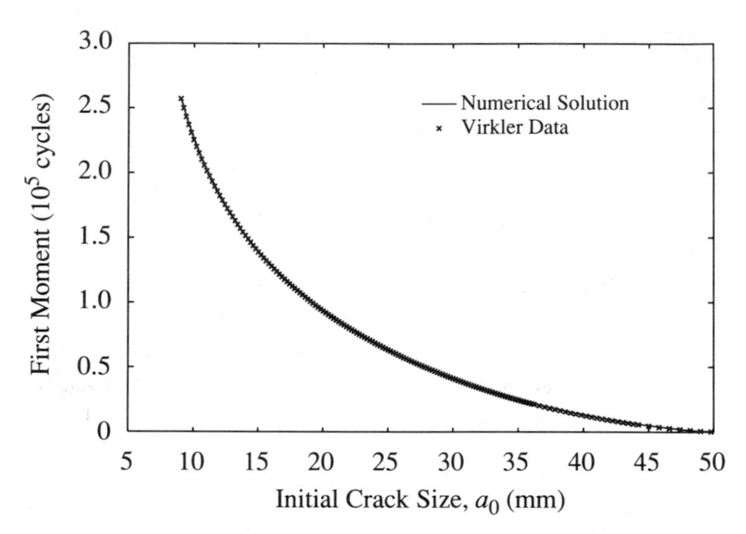

Figure 6.6. Comparison between the predicted mean time to reach $a_c = 49.8$ mm and the experimental data.

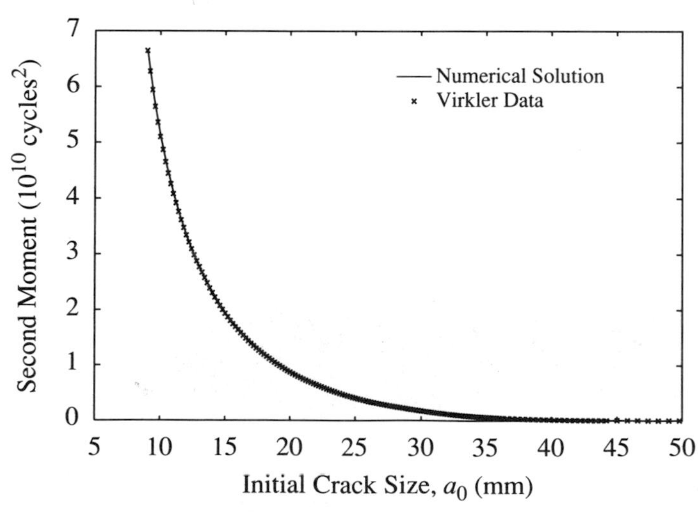

Figure 6.7. Comparison between the predicted mean-square time to reach $a_c = 49.8$ mm and the experimental data.

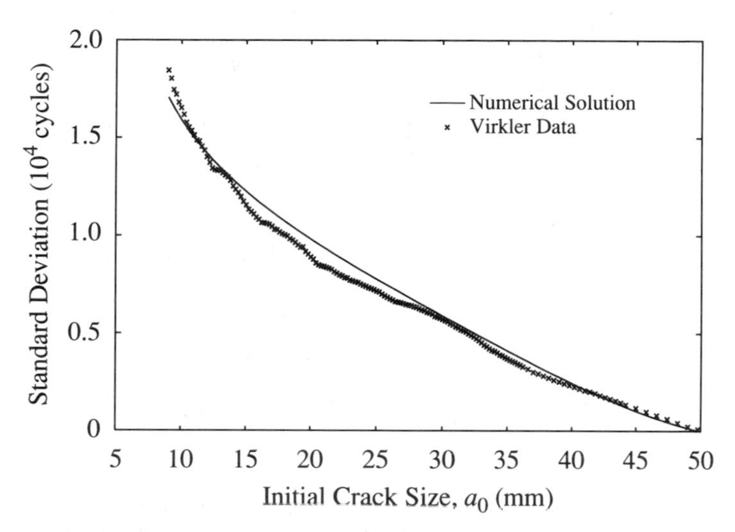

Figure 6.8. Comparison between the predicted standard deviation of the time to reach $a_c = 49.8$ mm and the experimental data.

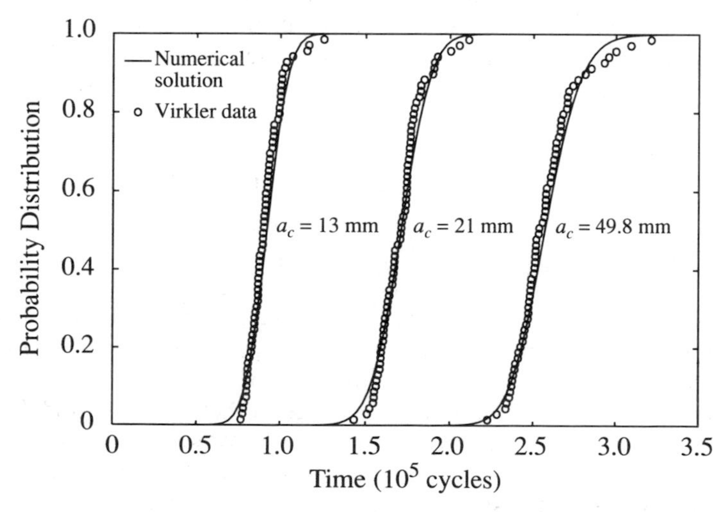

Figure 6.9. Comparison between the predicted distribution function of the time to reach a_c = 13, 21, and 49.8 mm and the experimental data (a_0 = 9 mm).

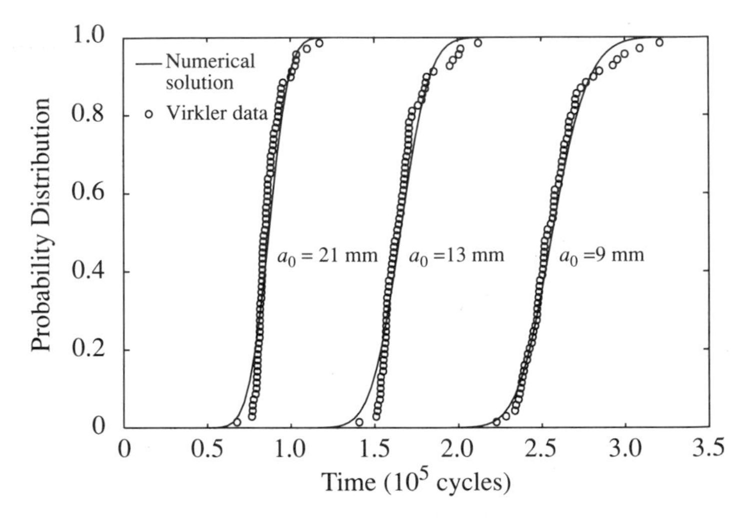

Figure 6.10. Comparison between the predicted distribution function of the time to reach $a_c = 49.8$ mm and the experimental data ($a_0 = 9$, 13, and 21 mm).

6.9, the initial crack size is held constant and the distribution of times to reach several crack sizes is presented. The distribution functions for several initial crack sizes, holding the critical crack size, are presented in Fig. 6.10. In all cases, the results fit the data very well, and no bias is observed.

Because of the excellent agreement of the results predicted by the model and the data, we conclude that the two-state crack growth model is capable of accurately describing the stochastic nature of the fatigue crack growth phenomenon for a wide variety of initial crack lengths, as well as critical crack lengths. Comparison of the two-state model with the Ghonem and Dore data [15] for several stress ratios is given in [194].

33. Multi-State Models for Random Loading

In Section 26, the physical aspects of crack retardation were discussed, and a phenomenological random birth model was presented that possessed many of the characteristics seen in empirical data. In this section, we present a recently developed multi-state fatigue crack growth model that includes load sequence effects through the introduction of a history-dependent parameter, the *reset stress*. The reset stress provides an indication of the state of residual compression at the crack tip. (See Veers [197], Veers et al. [198], and Winterstein and Veers [199].)

33.1 General Properties

Consider the Paris–Erdogan-type crack growth law modified for random loading and given by

$$\frac{dA}{dt} = C\,(\Delta K_{eff})^{m}, \tag{6.116}$$

where ΔK_{eff} is the effective stress intensity factor range defined as in the crack closure model of Elber (see Section 4.3), i.e.,

$$\Delta K_{eff} = B\,(a)\,\Delta S_{eff}\,\sqrt{\pi a}, \tag{6.117}$$

where $B\,(a)$ is the geometry factor and ΔS_{eff} is the effective stress intensity range. The basic idea here is that the crack will propagate only when the stress is great enough to cause the crack faces to fully separate or open. The effective stress range is then defined as

$$\Delta S_{eff} = \begin{cases} S_{\max} - S_{op}, & S_{\max} > S_{op},\ S_{\min} < S_{op} \\ S_{\max} - S_{\min}, & S_{\max} > S_{op},\ S_{\min} \ge S_{op} \\ 0, & S_{\max} \le S_{op}, \end{cases} \tag{6.118}$$

where S_{op} is the opening stress, which is determined experimentally. The effective stress intensity factor range then becomes

$$\Delta K_{eff} = \begin{cases} B\,(a)\,\sqrt{\pi a}\,(S_{\max} - S_{op})\,, & S_{\max} > S_{op},\ S_{\min} < S_{op} \\ B\,(a)\,\sqrt{\pi a}\,(S_{\max} - S_{\min}), & S_{\max} > S_{op},\ S_{\min} \ge S_{op} \\ 0, & S_{\max} \le S_{op}. \end{cases} \tag{6.119}$$

In the case of constant-amplitude loading, Elber proposed that the opening stress is related to the maximum stress as

$$S_{op} = c_f\,S_{\max}\,, \tag{6.120}$$

where c_f is usually considered to be a function of the stress ratio R.

For random loading, Eq. (6.120) is generalized as

$$S_{op} = c_f\,S_r, \tag{6.121}$$

where S_r, termed the *reset stress*, is defined as the stress level necessary to *reset* the maximum extent of the overload-affected zone at the crack tip. In this manner, sequence effects are included and the magnitude of the crack growth rate retardation is correlated with the magnitude of the overload.

To better understand the concept of reset stress, let us introduce a few quantities. A tensile overload applied to a cracked specimen is designated S_{ol} and the associated stress intensity factor is given by

$$K_{ol} = B(a)\sqrt{\pi a}\, S_{ol}. \tag{6.122}$$

Here, the maximum extent of the influence of the overload d_{ol} is approximated as twice the radius of the plastic zone (see Johnson [173]) and given by

$$d_{ol} = \frac{K_{ol}^2}{\lambda \pi S_y^2}, \tag{6.123}$$

where λ is a constraint factor ($\lambda = 1$ for plane stress and $\lambda = 3$ for plane strain), and S_y is the yield stress. The reset stress S_r can then be approximated as

$$S_r = \frac{S_y}{B(a)}\left(\frac{\lambda d_p}{a}\right)^{1/2}, \tag{6.124}$$

where d_p is defined as the distance from the crack tip to the maximum extent of the overload influence zone. Figure 6.11 provides a graphical depic-

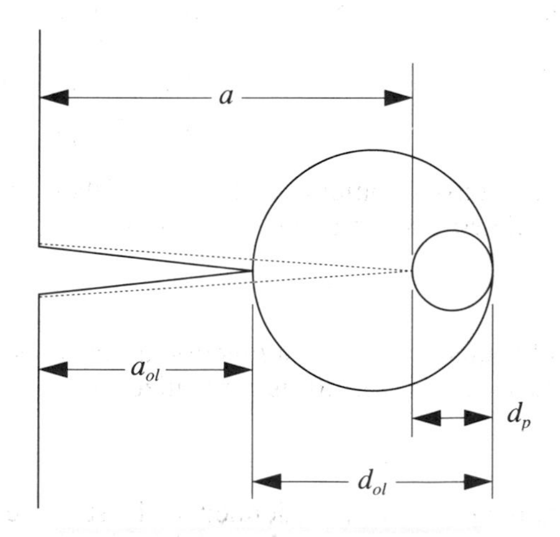

Figure 6.11. Overload-affected zone.

tion of the quantity d_p and the concept of the reset stress; the reset stress is the stress necessary to create an overload influence zone that has a magnitude greater than d_p.

33.2 Four-State Diffusion Model

To develop a model of crack growth that is a vectored diffusion process, we need to have an expression for the time derivative of the reset stress. Thus, taking the derivative of S_r with respect to time yields

$$\frac{dS_r}{dt} = \frac{\partial S_r}{\partial a}\frac{da}{dt} + \dot{S}H(\dot{S})H(S - S_r), \tag{6.125}$$

where S and $\dot{S}$ are the applied stress and its time derivative, respectively, and the $H(\cdot)$ is the Heaviside function. (See Eq. (5.108).) The first term in Eq. (6.125) is due to the gradual reduction in the reset stress as the crack grows through the plastic zone, and the second term is due to the instantaneous increase in the reset stress when the applied stress exceeds the reset stress. Substituting appropriate terms into Eq. (6.125) yields

$$\frac{dS_r}{dt} = -\frac{S_r}{B(a)}\left[\frac{dB(a)}{da} + \frac{1}{2a}\left(B(a) + \frac{\lambda S_y^2}{B(a)S_r^2}\right)\right]\frac{da}{dt} + \dot{S}H(\dot{S})H(S - S_r). \tag{6.126}$$

Upon examination of the previous expression, we see that the applied stress and stress rate must be states variables in the model. The applied stress can be modelled as a white noise passed through a linear second-order filter. Therefore, we define the state vector for this model as $\mathbf{X} = [A, S_r, S, \dot{S}]$, which is governed by the vector stochastic differential equation,

$$\dot{\mathbf{X}} = \boldsymbol{\eta}(\mathbf{X}) + \mathbf{D}\xi(t, \gamma), \tag{6.127}$$

where $\boldsymbol{\eta}$ is the drift vector and $\mathbf{D}$ is the diffusivity matrix. The drift vector can be written explicitly as

$$\boldsymbol{\eta} = [\eta_A, \eta_{S_r}, \eta_S, \eta_{\dot{S}}]^T, \tag{6.128}$$

where

$$\eta_A = C[B(A)\sqrt{\pi A}\Delta S_{eff}]^m, \tag{6.129a}$$

$$\eta_{S_r} = -\frac{S_r}{B(a)}\left[\frac{dB(a)}{da} + \frac{1}{2a}\left(B(a) + \frac{\lambda S_y^2}{B(a) S_r^2}\right)\right]\eta_A \qquad (6.129b)$$
$$+ \dot{S} H(\dot{S}) H(S - S_r),$$

$$\eta_S = \dot{S}, \qquad (6.129c)$$

$$\eta_{\dot{S}} = -\omega_0^2 (S - \mu_S) - 2\zeta\omega_0 \dot{S}, \qquad (6.129d)$$

μ_S is the mean of the applied stress process, and the parameters ω_0 and ζ describe the characteristics of the linear filter for the applied stress process. In addition, $D_{44} = 4\zeta\omega_0^3\sigma_S^2$ is the only nonzero element in the diffusivity matrix where σ_S^2 is the stationary variance of the applied stress process. To complete the formulation of this four-state model, S_{max} must be approximated in the expression for ΔS_{eff}, Eq. (6.118). This is accomplished by modelling the maximum of the applied stress as

$$S_{max} = \mu_S + \mathcal{A}, \qquad (6.130)$$

where $\mathcal{A}$ is the amplitude envelope given in terms of the states by

$$\mathcal{A} = \sqrt{(S - \mu_S)^2 + (\dot{S}/\omega_0)^2}. \qquad (6.131)$$

33.3 Reliability Problem Formulation

In this section, we develop an initial-boundary value problem for the probability that the crack length is less than the critical crack length based on the four-state model. (See Enneking [167].) The transition probability density function $p = p(\mathbf{x}, t | \mathbf{x}_0, t_0)$ for the four-state vector $\mathbf{x} = [A, S_r, S, \dot{S}]$ is governed by the Fokker–Planck–Kolmogorov equation,

$$\frac{\partial p}{\partial t} = -\frac{\partial}{\partial a}[\eta_A(\mathbf{x}) p] - \frac{\partial}{\partial s_r}[\eta_{S_r}(\mathbf{x}) p] - \frac{\partial}{\partial s}[\eta_S(\mathbf{x}) p]$$
$$-\frac{\partial}{\partial \dot{s}}[\eta_{\dot{S}}(\mathbf{x}) p] + 2\zeta\omega_0^3\sigma_S^2 \frac{\partial^2 p}{\partial \dot{s}^2}. \qquad (6.132)$$

The backward Kolmogorov equation also governs the probability density function and is given by

$$\frac{\partial p}{\partial t} = \eta_A(\mathbf{x}_0)\frac{\partial p}{\partial a_0} + \eta_{S_r}(\mathbf{x}_0)\frac{\partial p}{\partial s_{r0}} + \eta_S(\mathbf{x}_0)\frac{\partial p}{\partial s_0} + \eta_{\dot{S}}(\mathbf{x}_0)\frac{\partial p}{\partial \dot{s}_0} + 2\zeta\omega_0^3\sigma_S^2\frac{\partial^2 p}{\partial \dot{s}_0^2}\,. \tag{6.133}$$

In a development similar to that in Section 32.1, we assume that fatigue failure has not occurred if the crack has not reached a length of a_c. In addition, we define failure of the structure when the applied stress reaches the allowable stress, s_{all}. The cumulative distribution of times to reach a given crack size a_c, or conversely, the probability that a crack will not reach a critical crack size in a given time, is then defined as (conditional on the initial value of the state vector $\mathbf{x}_0$)

$$R = R(t|\mathbf{x}_0) = P\{A(t) < a_c, S < s_{all}; t|\mathbf{x}_0; t_0\} = \int_{a_0}^{a_c}\int_{0}^{s_{all}}\int_{-\infty}^{s_r}\int_{-\infty}^{\infty} p(a, s_r, s, \dot{s}, t|\,\mathbf{x}_0; t_0)\,d\dot{s}\;ds\;ds_r da\,. \tag{6.134}$$

The reliability function R also satisfies the backward Kolmogorov equation (6.133), i.e.,

$$\frac{\partial R}{\partial t} = \eta_A(\mathbf{x}_0)\frac{\partial R}{\partial a_0} + \eta_{S_r}(\mathbf{x}_0)\frac{\partial R}{\partial s_{r0}} + \eta_S(\mathbf{x}_0)\frac{\partial R}{\partial s_0} + \eta_{\dot{S}}(\mathbf{x}_0)\frac{\partial R}{\partial \dot{s}_0} + 2\zeta\omega_0^3\sigma_S^2\frac{\partial^2 p}{\partial \dot{s}_0^2}\,. \tag{6.135}$$

Only one boundary condition needs to be specified in the a_0, s_{r0}, and s_0 directions for the initial-boundary value problem to be well-posed. (See Fichera [170].) Possible initial and boundary conditions are suggested on physical grounds and given by

$$R(0|a_c, s_{r0}, s_0, \dot{s}_0) = 1, \qquad 0 \le a_0 < a_c,\ s_0 \le s_{r0} < s_{all}, \tag{6.136}$$

$$R(t|a_c, s_{r0}, s_0, \dot{s}_0) = 0, \qquad s_0 \le s_{r0} < s_{all}, \tag{6.137}$$

$$R(t|a_0, s_{all}, s_0, \dot{s}_0) = 0, \qquad 0 \le a_0 \le a_c,\ s_0 < s_{all}, \tag{6.138}$$

$$\lim_{s_0 \to -\infty} R(t|a_0, s_{r0}, s_0, \dot{s}_0) \rightarrow 0, \qquad 0 \le a_0 \le a_c,\ s_{r0} < s_{all}, \tag{6.139}$$

$$\lim_{|\dot{s}_0| \to \infty} R(t | a_0, s_{r0}, s_0, \dot{s}_0) \to 0, \qquad 0 \leq a_0 \leq a_c, s_0 \leq s_{r0} < s_{all}. \quad (6.140)$$

The determination of the preceding boundary conditions can be made following logic similar to that presented for the two-state crack growth problem in Section 32.1. First, we note that the initial stress may not be greater than the initial reset stress and that both must be less than the maximum allowable stress, i.e., $s_0 \leq s_{r0} \leq s_{all}$. In addition, the reset stress must be greater than zero, i.e., $s_{r0} \geq 0$. The initial condition is given by Eq. (6.136) and simply states that the reliability is one at $t = 0$ as long as the initial crack length is less than the critical crack length. The boundary condition given in Eq. (6.137) indicates that the reliability is zero if the initial crack size is the critical crack size. Equation (6.138) states that the reliability should be zero when the initial reset stress is equal to the maximum allowable stress, σ_{all}. The boundary condition in Eq. (6.139) indicates that as the initial stress goes to negative infinity, the stress amplitude will become very large, resulting in rapid crack growth, system failure, and zero reliability. The last boundary condition (6.140) states that the reliability will go to zero as the initial stress rate and, thus, the stress amplitude (e.g., see Eq. (6.131)) go to infinity, again causing the crack to grow rapidly and the system to fail.

The boundary value for the statistical moments of time to reach the critical crack size can also be formulated in a manner similar to that found in Section 32.2. This problem may be somewhat simpler than the reliability problem to solve, since it is not time-dependent.

The model presented in this section is physically appealing, and Monte Carlo simulation indicates that many important features of the fatigue phenomena under random loading are captured. (See [197].) Unfortunately, there are no analytical solutions, and more research is needed to develop numerical solutions for this problem formulation.

To alleviate this problem, analytical results for the mean life have been developed that use a set of scalar diffusion models that are on different time scales; e.g., models of S_r at various crack sizes. These models are shown to produce results that agree well with simulation. (See [197] and [199].)

34. Other Approaches

A number of researchers have suggested that the multiplicative random process in Eqs. (6.35) and (6.63) be a function of crack size a, rather than time or cycle number. For constant-amplitude tests in tightly controlled laboratory settings, this is physically reasonable, as the majority of the variability found in fatigue crack growth is attributed to the inhomogeneity of the material, i.e., the crack growth rate varies as a function of the resistance of the material through which it is propagating. In these approaches, the randomness in crack growth rates is broken down into (i) specimen-to-specimen variation, and (ii) within-specimen variation. The specimen-to-specimen variation is represented by random variables, whereas the within-specimen variation is modelled using a random process. (See Ortiz and Kiremidjian [182], [183], Madsen et al. [31], Ditlevsen [162], and Ditlevsen and Olesen [163].)

Consider again the Paris–Erdogan equation given by

$$\frac{da}{dN} = C(\Delta K)^m, \tag{6.141}$$

where N is the number of cycles. The material parameter m is often assumed to be constant, and the parameter C randomized, as (see Madsen et al. [31])

$$C = C(a) = \frac{C_1}{C_2(a)}. \tag{6.142}$$

In this approach, C_1 is a positive random variable describing fluctuations between the mean behavior of different specimens, whereas $C_2(a)$ is a positive random process modelling the deviation of crack growth from the mean crack path within each specimen. The process $C_2(a)$ is assumed to be homogeneous and lognormally distributed with a mean value of one. Substituting Eq. (1.22) for ΔK, Eq. (6.141) can be rewritten as

$$\frac{dA}{dN} = \frac{C_1}{C_2(a)}\left[B(a)\,\Delta S\sqrt{\pi a}\right]^m, \tag{6.143}$$

where $B(a)$ is the geometry function, and ΔS is the stress range. For constant-amplitude loading, we can separate and integrate Eq. (6.143) to obtain

$$\Psi(a) = C_1(\Delta S)^m N, \tag{6.144}$$

where $\Psi(a)$ is a random function, denoted as the damage function, given by

$$\Psi(a) = \int_{a_0}^{a} C_2(z)\frac{dz}{B(z)^m (z\pi)^{m/2}}. \tag{6.145}$$

The probability distribution of $\Psi(a)$, defined by the stochastic integral given in the previous equation, is not readily determined. However, the conditional mean and the covariance function of damage can be given respectively by

$$\langle \Psi(a) | a_0, B, m \rangle = \int_{a_0}^{a} \frac{\langle C_2(z) \rangle}{B(z)^m (z\pi)^{m/2}} dz = \int_{a_0}^{a} \frac{1}{B(z)^m (z\pi)^{m/2}} dz , \tag{6.146}$$

$$\begin{aligned} K_\Psi(a_1, a_2 | a_0, B, m) &= \mathrm{Cov}[\Psi(a_1), \Psi(a_2) | a_0, B, m] \\ &= \int_{a_0}^{a_1}\int_{a_0}^{a_2} \frac{K_{C_2}(z_1, z_2)}{B(z_1)^m (z_1\pi)^{m/2} B(z_1)^m (z_2\pi)^{m/2}} dz_1 dz_2, \end{aligned} \tag{6.147}$$

where $K_{C_2}(z_1, z_2)$ is the covariance function for the random process $C_2(a)$.

Based upon experimental results [182], the correlation function for $C_2(a)$ is found to rapidly decrease to zero, and therefore, it may be reasonably approximated as a white noise process. Noting that the function $\Psi(a)$ is the weighted integral of a random process with small correlation, then when $a - a_0$ is relatively large (i.e., long fatigue lives), the distribution of $\Psi(a)$ is well-approximated by a normal distribution with mean and variance determined from Eqs. (6.146)–(6.147).

If the failure of the specimen is defined as the point at which the crack size reaches the critical crack size a_c, then the probability of failure is given by

$$\begin{aligned} &P\{\Psi(a_c) \geq \Psi(a)\} \\ &= P\left\{ \int_{a_0}^{a} C_2(z) \frac{dz}{B(z)^m (z\pi)^{m/2}} \geq C_1 (\Delta S)^m N \right\}. \end{aligned} \tag{6.148}$$

While these probabilities are not easily determined analytically, they can be effectively calculated by first-order or second-order reliability methods using computer programs such as PROBAN [185].

This approach can also be extended to include variable-amplitude loading when sequence effects are neglected. The differential equation for crack growth can be written in an incremental form given by

$$\Delta A_i = \frac{C_1}{C_2(A)} B(A)^m (\Delta S_i)^m (A_i \pi)^{m/2}, \tag{6.149}$$

where ΔA_i is the increase in the crack length due to the ith applied stress cycle, and A_i is the crack length in the ith stress cycle. Equation (6.144) then becomes

$$\Psi(a) = C_1 \sum_{i=1}^{N} \Delta S_i^m. \tag{6.150}$$

The probability that the crack size reaches the critical crack size a_c is then given by

$$P\{\Psi(a_c) \geq \Psi(a)\}$$

$$= P\left\{\int_{a_0}^{a} C_2(z) \frac{dz}{B(z)^m (z\pi)^{m/2}} > C_1 \sum_{i=1}^{N} \Delta S_i^m\right\}. \tag{6.151}$$

This model has been extensively employed in inspection and maintenance problems for offshore structures [179].

Chapter VII

Comparisons and Applications

35. Introductory Remarks

Stochastic modelling and analysis of fatigue stimulates a number of additional questions that are not intended to be a subject of detailed consideration in this book. Nevertheless, we wish to mention some problems concerned, generally speaking, with model validation and possible applications.

The question of whether a given model satisfactorily represents a real phenomenon (and available empirical information) remains an important and intriguing one. However, the validity of a model can only be specified in relative terms. We can only say that a model is satisfactory if it meets the appropriate criteria. To validate a model, we have to *compare* its predictions with the corresponding characteristics of empirical data. Also, the comparisons between various models are often of interest. In performing fatigue experiments under random loading, various random signals can be used, and therefore, the characterization of their divergence can constitute an important issue.

In all the situations mentioned previously, there is a need for appropriate measures of divergence or diversity between a model and observation, between two models, or between two random signals. Such measures (known also as *distance measures*) were introduced in various contexts, e.g., in signal processing, in pattern recognition, and in mathematical statistics where comparisons between theoretical and empirical probability distributions have been performed for many years. Although the measures of divergence can be defined in different ways, we wish to emphasize the importance of the *theoretic–information* approach. (See Sobczyk [213].)

It is known that the concepts of entropy and mutual information (defined in Section 9.4) are extremely useful and effective for evaluating the performance of communication systems. This high effectiveness is, most likely, a reason that information theory is often considered synonymous with communication theory. Such a narrow meaning of information theory cannot be justified, since information theory is, in essence, a branch of probability theory and mathematical statistics, and as such, should be applicable to analysis of any physical or engineering system. Indeed, the theoretic–information approach is well-known in physics, especially in thermodynamics where the relationship between the amount of information on a physical system and its thermodynamical entropy is well-established. (See Poplavsky [212].)

Information theory can be especially relevant to data processing, statistical inference, and, of particular importance here, model validation. In fact, information in a technically defined sense was first introduced in statistics by R. A. Fisher in 1925, in his work on the theory of estimation. His concept of a measure of the amount of information supplied by data about an unknown parameter is well-known to statisticians. In general, the apparatus of information theory is applicable to any probabilistic system of observations, since whenever we make observations (or design and conduct experiments), we seek information. The question that arises in the process of model building on the basis of empirical data is: How much information can we infer from a particular set of observations or experiments about the sampled phenomenon? A particular problem concerns estimation of an unobserved quantity X through observations on another quantity Y; these quantities can be random variables or random processes. The concepts of entropy and mutual information (between random variables), briefly presented in Section 9.4, can be used in many different situations.

In this chapter, we wish to provide some formulae originating in research in information theory that can be useful in comparisons of various probabilistic models of fatigue and in model validation. In the last section, an application of stochastic analysis to real engineering problems will be given.

36. Distance Measures between Probability Distributions

As seen in previous chapters, stochastic models of the fatigue process yield different probability distributions of a crack size and life of a specimen. To be able to compare the predictions of two models, as well as to conduct comparisons of the specific model with empirical data, one asks how different are, or what is the distance between, corresponding probability distributions?

A distance or divergence between two probability distributions can be measured in various ways. However, all distance measures must possess some basic properties that are *natural* for such measures. Let us assume that we have two probability distributions, F_1 and F_2, which are absolutely continuous; that is, they possess densities $f_1(x)$ and $f_2(x)$. A general class of distance measures can be defined as (see Csiszar [204])

$$d(F_1, F_2) = h_1\left(\left\langle h_2\left(\frac{f_2}{f_1}\right)\right\rangle\right), \tag{7.1}$$

where h_2 is a continuous convex real function on $R_+ = [0, \infty)$, h_1 is an increasing function on $R_1 = (-\infty, +\infty)$, and $\langle\cdot\rangle$ is the expected value with respect to distribution F_1; the quantity,

$$\varphi = \frac{f_2(x)}{f_1(x)}, \tag{7.2}$$

is often called the *likelihood ratio*.

Let us notice that when we take h_1 to be identity, then Eq. (7.1) takes the form

$$d(F_1, F_2) = \int_{-\infty}^{\infty} h_2\left(\frac{f_2(x)}{f_1(x)}\right) f_1(x)\,dx, \tag{7.3}$$

with $d(F_1, F_2) = 1$ if and only if $f_1(x) = f_2(x)$. There are several particular cases of Eq. (7.3) worth mentioning:

Variational distance: $h_1(x) = \frac{1}{2}x$, $h_2(x) = |1 - x|$,

$$d(F_1, F_2) = \frac{1}{2}\int_{-\infty}^{\infty} |f_1(x) - f_2(x)|\,dx = \mathcal{V}(F_1, F_2). \tag{7.4}$$

Hellinger distance: $h_1(x) = \frac{1}{2}x$, $h_2(x) = (\sqrt{x} - 1)^2$,

$$d(F_1, F_2) = \frac{1}{2}\int_{-\infty}^{\infty} [\sqrt{f_2(x)} - \sqrt{f_1(x)}]^2 dx = \mathcal{H}(F, F_2). \tag{7.5}$$

Kullback–Leibler information: $h_1(x) = x,\ h_2(x) = -\log x,$

$$d(F_1, F_2) = \int_{-\infty}^{\infty} f_1(x) \log \frac{f_1(x)}{f_2(x)} dx = \mathcal{J}(F_1, F_2). \tag{7.6}$$

Kullback–Leibler divergence: $h_1(x) = x,\ h_2(x) = (x-1)\log x,$

$$d(F_1, F_2) = \int_{-\infty}^{\infty} [f_2(x) - f_1(x)] \log \frac{f_2(x)}{f_1(x)} dx = \mathcal{K}(F_1, F_2). \tag{7.7}$$

It is seen that $\mathcal{K}(F_1, F_2) = \mathcal{J}(F_1, F_2) + \mathcal{J}(F_2, F_1)$.

Bhattacharrya distance: $h_1(x) = -\log(-x),\ h_2(x) = -\sqrt{x};\ x > 0,$

$$d(F_1, F_2) = -\log d_1(F_1, F_2) = \mathcal{B}(F_1, F_2), \tag{7.8}$$

where

$$d_1(F_1, F_2) = \int \sqrt{f_1(x) f_2(x)}\, dx = 1 - \mathcal{H}^2(F_1, F_2). \tag{7.9}$$

The distance measures in Eqs. (7.3)–(7.8) are commonly used in various problems in information theory and pattern recognition. However, there also exist other distance measures that are not special cases of Eq. (7.1); for example,

$$\rho(F_1, F_2) = [\int \{f_1(x) - f_2(x)\}^2]^{1/2} dx \tag{7.10}$$

or

$$\rho(F_1, F_2) = \int |f_1(x) - f_2(x)|^{\alpha} dx, \qquad \alpha > 0, \alpha \neq 1. \tag{7.11}$$

There is an interesting relationship between the Shanon information measure (or Shanon mutual information) $I(X, Y)$ defined by Eq. (2.81) and the Kullback–Leibler information measure (7.6). To see this relation, let us assume that we wish to identify an unobservable variable $X(\gamma)$ on the basis of observation of another variable $Y(\gamma)$ that is statistically related to $X(\gamma)$. The Shanon measure of information regarding the true value of $X(\gamma)$ when $Y(\gamma)$ is observed is given by Eq. (2.81). Consider the two–dimensional analog of Eq. (7.6) given by

$$\mathcal{J}(f_1,f_2) = \mathcal{J}[f_{XY}(x,y), f_X(x) f_Y(y)] = I(X,Y)$$
$$= \int_{-\infty}^{\infty}\int_{-\infty}^{\infty} f(x,y) \log \frac{f(x,y)}{f_X(x) f_Y(y)} dx dy, \tag{7.12}$$

where $f_1 = f_{XY}(x,y)$ and $f_2 = f_X(x) f_Y(y)$. It is clear from Eqs. (2.81) and (7.12) that the Shanon amount of information regarding $X(\gamma)$ provided by observation of $Y(\gamma)$ can be considered to be equal to the Kullback–Leibler information between $f_X(x) f_Y(y)$ and the joint distribution $f_{XY}(x,y)$.

Another quantity related to $I(X,Y)$ and $\mathcal{J}$ is the so-called *inaccuracy measure* introduced by Kerridge [207],

$$\mathcal{A}(f_1,f_2) = -\int_{-\infty}^{\infty} f_1(x) \log f_2(x)\, dx. \tag{7.13}$$

Using the definitions of entropy in Eq. (2.71), the Kullback–Leibler information, and the inaccuracy measure in Eq. (7.13), we come to the following relationship:

$$\mathcal{A}(f_X,f_Y) = -\int_{-\infty}^{\infty} f_X(x) \log f_X(x)\, dx + \int_{-\infty}^{\infty} f_X(x) \log \frac{f_X(x)}{f_Y(x)} dx$$
$$= H[f_X(x)] + \mathcal{J}(f_X,f_Y), \tag{7.14}$$

where $H[f_X(x)]$ denotes the entropy of the distribution $f_X(x)$. The inaccuracy (7.13) is a nonnegative and additive quantity. If $f_X(x)$ represents the true probability distribution of a random variable $X(\gamma)$ — which can be, for instance, a crack size at a fixed time t — and $f_Y(x)$ is an approximation of $f_X(x)$ based on some inaccurate knowledge of $X(\gamma)$, then it follows from Eq. (7.14) that the inaccuracy measure $\mathcal{A}(f_X,f_Y)$ can be regarded as a measure of total uncertainty of $X(\gamma)$ that occurs due to its inherent randomness (entropy of $X(\gamma)$) and because of the inaccuracy of our knowledge of the true probability distribution.

37. Spectral Distance Measures

Here, we will discuss briefly the distance measures between stochastic processes based on their second-order properties expressed by spectral densities. Such measures can be useful in characterization of divergence between various random load processes.

Let $X_1(t)$ and $X_2(t)$ be two stationary stochastic processes with joint n–dimensional probability densities $f_{X_1}(x_1, \ldots, x_n)$ and $f_{X_2}(x_1, \ldots, x_n)$, respectively; these densities correspond to the time instants $t_1, \ldots, t_n$. Let us denote $\mathbf{x}_n = (x_1, \ldots, x_n)^T$.

The *Kullback–Leibler information rate* between $X_1(t)$ and $X_2(t)$ is defined as

$$\mathcal{I} = \lim_{n \to \infty} \frac{1}{n} \mathcal{I}_n, \tag{7.15}$$

where $\mathcal{I}_n$ is the nth-order Kullback–Leibler information measure defined by (see Kullback [208])

$$\mathcal{I}_n = \int f_{X_1}(\mathbf{x}_n) \ln \frac{f_{X_1}(\mathbf{x}_n)}{f_{X_2}(\mathbf{x}_n)} d\mathbf{x}_n. \tag{7.16}$$

If $X_1(t)$ and $X_2(t)$ are stationary Gaussian processes, then

$$\begin{aligned} \mathcal{I}_n = {} & \frac{1}{2} \ln \left| \frac{\mathbf{K}_2}{\mathbf{K}_1} \right| + \frac{1}{2} \operatorname{tr}\left[\mathbf{K}_1 \left(\mathbf{K}_2^{-1} - \mathbf{K}_1^{-1}\right)\right] \\ & + \frac{1}{2} \operatorname{tr}\left[\mathbf{K}_2^{-1} \left(m_{X_1} - m_{X_2}\right) \left(m_{X_1} - m_{X_2}\right)^T\right], \end{aligned} \tag{7.17}$$

where $\mathbf{K}_1$ and $\mathbf{K}_2$ are the covariance matrices and m_{X_1} and m_{X_2} are the mean values (vectors) of X_1 and X_2, respectively. If the means are equal, i.e., $m_{X_1} = m_{X_2}$, then (see Kazrakos and Papontoni-Kazakos [206])

$$\mathcal{I} = \frac{1}{4\pi} \int_{-\pi}^{\pi} \left(\ln \frac{g_{X_2}(\omega)}{g_{X_1}(\omega)} + \frac{g_{X_1}(\omega)}{g_{X_2}(\omega)} - 1 \right) d\omega, \tag{7.18}$$

where $g_{X_1}(\omega)$ and $g_{X_2}(\omega)$ are the spectral densities of $X_1(t)$ and $X_2(t)$, respectively.

Another quantity that has turned out to be useful in system identification and in mathematical statistics is the *Fisher information matrix* $\mathfrak{F}$ for a random process $X(t)$ with parametric probability density $f_X(\mathbf{x}_n | \boldsymbol{\theta})$, where $\boldsymbol{\theta} = (\theta_1, \ldots, \theta_l)^T$. This matrix is defined as

$$\mathfrak{F}(\boldsymbol{\theta}) = \lim_{n \to \infty} \frac{1}{n} \mathfrak{F}_n(\boldsymbol{\theta}), \tag{7.19}$$

where $\mathfrak{F}_n(\boldsymbol{\theta})$ is the $(l \times l)$ matrix whose (i, k)th element for $i, k = 1, 2, \ldots, l$ is given by

$$\begin{aligned} \{\mathfrak{F}_n(\boldsymbol{\theta})\}_{i,k} &= -\int \frac{\partial^2 \ln f_X(\mathbf{x}_n | \boldsymbol{\theta})}{\partial \theta_i \partial \theta_k} f(\mathbf{x}_n | \boldsymbol{\theta})\, d\mathbf{x}_n \\ &= \int \frac{\partial \ln f_X(\mathbf{x}_n | \boldsymbol{\theta})}{\partial \theta_i} \frac{\partial \ln f_X(\mathbf{x}_n | \boldsymbol{\theta})}{\partial \theta_k} d\mathbf{x}_n. \end{aligned} \tag{7.20}$$

For stationary Gaussian processes, the elements of $\mathfrak{F}(\boldsymbol{\theta})$ are (see [215])

$$\{\mathfrak{F}(\boldsymbol{\theta})\}_{i,k} = \frac{1}{4\pi} \int_{-\pi}^{\pi} \left[\frac{1}{g_X^2(\omega;\boldsymbol{\theta})} \frac{\partial g_X(\omega;\boldsymbol{\theta})}{\partial \theta_i} \frac{\partial g_X(\omega;\boldsymbol{\theta})}{\partial \theta_k} \right] d\omega. \tag{7.21}$$

A number of other theoretic–information measures are related to the Fisher information matrix. This relationship can be established using the Taylor expansion up to the second order.

Consider the Kullback–Leibler information rate for a random process with spectra $g(\omega;\boldsymbol{\theta})$ and $g(\omega;\underset{\sim}{\boldsymbol{\theta}})$, respectively, where parameters $\boldsymbol{\theta}$ and $\underset{\sim}{\boldsymbol{\theta}}$ are l–dimensional vectors; hence, by virtue of Eq. (7.18),

$$\mathcal{J}(\boldsymbol{\theta}, \underset{\sim}{\boldsymbol{\theta}}) = \frac{1}{4\pi} \int_{-\pi}^{\pi} \left[\ln \frac{g(\omega;\underset{\sim}{\boldsymbol{\theta}})}{g(\omega;\boldsymbol{\theta})} + \frac{g(\omega;\boldsymbol{\theta})}{g(\omega;\underset{\sim}{\boldsymbol{\theta}})} - 1 \right] d\omega. \tag{7.22}$$

By applying the Taylor series expansion up to the second order at $\underset{\sim}{\boldsymbol{\theta}} = \boldsymbol{\theta}$, we have (see Sugimoto and Wada [214])

$$\begin{aligned} \mathcal{J}(\boldsymbol{\theta}, \underset{\sim}{\boldsymbol{\theta}}) \approx{}& \mathcal{J}(\boldsymbol{\theta}, \boldsymbol{\theta}) + \left(\frac{\partial \mathcal{J}(\boldsymbol{\theta}, \underset{\sim}{\boldsymbol{\theta}})}{\partial \underset{\sim}{\boldsymbol{\theta}}}\right)^T \Bigg|_{\underset{\sim}{\boldsymbol{\theta}} = \boldsymbol{\theta}} (\underset{\sim}{\boldsymbol{\theta}} - \boldsymbol{\theta}) \\ &+ \frac{1}{2} (\underset{\sim}{\boldsymbol{\theta}} - \boldsymbol{\theta})^T \frac{\partial^2 \mathcal{J}(\boldsymbol{\theta}, \underset{\sim}{\boldsymbol{\theta}})}{\partial \underset{\sim}{\boldsymbol{\theta}} \partial \underset{\sim}{\boldsymbol{\theta}}} \Bigg|_{\underset{\sim}{\boldsymbol{\theta}} = \boldsymbol{\theta}} (\underset{\sim}{\boldsymbol{\theta}} - \boldsymbol{\theta}), \end{aligned} \tag{7.23}$$

where

$$\mathcal{J}(\boldsymbol{\theta}, \boldsymbol{\theta}) = 0, \qquad \left.\frac{\partial \mathcal{J}(\boldsymbol{\theta}, \underset{\sim}{\boldsymbol{\theta}})}{\partial \underset{\sim}{\boldsymbol{\theta}}}\right|_{\underset{\sim}{\boldsymbol{\theta}} = \boldsymbol{\theta}} = 0, \tag{7.24}$$

and

$$\left.\frac{\partial^2 \mathcal{J}(\boldsymbol{\theta}, \underset{\sim}{\boldsymbol{\theta}})}{\partial \underset{\sim}{\boldsymbol{\theta}} \partial \underset{\sim}{\boldsymbol{\theta}}}\right|_{\underset{\sim}{\boldsymbol{\theta}} = \boldsymbol{\theta}} = \frac{1}{4\pi} \int_{-\pi}^{\pi} \frac{1}{g^2(\omega;\boldsymbol{\theta})} \left[\frac{\partial g(\omega;\boldsymbol{\theta})}{\partial \boldsymbol{\theta}}\right] \left[\frac{\partial g(\omega;\boldsymbol{\theta})}{\partial \boldsymbol{\theta}}\right]^T d\omega . \tag{7.25}$$

Therefore, from Eq. (7.21), we have the approximation,

$$\mathcal{J}(\boldsymbol{\theta}, \underset{\sim}{\boldsymbol{\theta}}) \approx \frac{1}{2} (\underset{\sim}{\boldsymbol{\theta}} - \boldsymbol{\theta})^T \mathfrak{F}(\theta) (\underset{\sim}{\boldsymbol{\theta}} - \boldsymbol{\theta}) . \tag{7.26}$$

The Kullback–Leibler information characterizes, in essence, an amount of information about one stationary process carried out in observation of the second one. If the process in question is Gaussian, this information can be expressed in terms of spectral densities. However, in some situations, one may wish to characterize a divergence between two spectral densities in a more *geometrical way*, treating spectra $g_1(\omega)$ and $g_2(\omega)$ as *points* of the appropriate metric space.

Let us assume that $g_1(\omega)$ and $g_2(\omega)$ belong to the space L_p of functions integrable with power $p > 1$. In such a case, a distance between g_1 and g_2 can be defined as

$$d(g_1, g_2) = \|g_1 - g_2\|_{L_p} = \left[\int |g_1(\omega) - g_2(\omega)|^p d\omega \right]^{1/p}, \tag{7.27}$$

which, for $p = 2$, coincides with the distance measure given in Eq. (7.10) between two probability densities.

The distances defined by norms (in the appropriate functional spaces) — hence, in particular, the distance in Eq. (7.27) — are *true* distances in a sense that they satisfy the symmetry property and the triangular inequality, i.e.,

$$\begin{aligned} d(g_1, g_2) &= d(g_2, g_1), \\ d(g_1 + g_2, 0) &\le d(g_1, 0) + d(g_2, 0) . \end{aligned} \tag{7.28}$$

The distance measure in Eq. (7.27) can be used to define other divergence measures between spectra; for example,

$$d_1(g_1, g_2) = d\left(1, \frac{g_2}{g_1}\right). \tag{7.29}$$

A measure that is probably the oldest one is defined by the L_p norm of the difference of logarithms of the spectra

$$d_2(g_1, g_2) = \|\log g_1 - \log g_2\|_{L_p} = \left\|\log \frac{g_1}{g_2}\right\|_{L_p}. \tag{7.30}$$

The most common choices are: $p = 1$ (mean absolute distance) and $p = 2$ (mean quadratic distance). These distances were first used in speech signal processing and are directly related to decibel variations in the log spectral domain.

38. Informational Quality of Models

Since the entropy (defined in Section 9.4) characterizes an informational content, prior to experiment, it can be used for judging of the model quality and for comparison of two models. Indeed, if $H_A(t)$ and $H_B(t)$ are the entropies of model A and B, then the quantity

$$d(t) = H_A(t) - H_B(t) \tag{7.31}$$

describes a difference between the informational capacity of model A and model B at each fixed time t. Using an integrated entropy, we get the following quantity:

$$d = \int_{t_0}^{t^*} [H_A(t) - H_B(t)]\, dt, \tag{7.32}$$

where t^* is a characteristic time instant; if $H(t)$ denotes the entropy of the fatigue crack growth process at time t, then t^* can be taken as t_{cr}. In Section 23.3, we indicated that the quantity (7.32) can be used for relating the parameters of a new model to the known parameters of another model.

In modelling of physical phenomena, i.e., particularly in modelling of the fatigue process, one wishes to find a model that would be the best among all admissible models. This requires introducing an appropriate quality criterion. A quality of a model is characterized by a functional $Q(Y, Y_M)$ defined on the output of the real system Y (in our case: real fatigue crack growth) and on the model output Y_M; the quality criterion selects from all admissible models such a model for which the functional

$Q(Y, Y_M)$ takes an extremal value. This criterion expresses how close the model should be to the true physical process, and it is a key issue in modelling. Most often, the root mean-square error (between Y and Y_M) is taken as the quality criterion. (See [205].)

It seems, however, that a more general quality criterion could also be useful; we mean, for example, criteria based on informational content of the model. A criterion of this type can be formulated as: a model M_0 should be selected from the class $\{M\}$ of the admissible models in such a way that the model output Y_M contains maximum information (in the Shannon sense) about the real system output Y, that is,

$$\begin{aligned} Q(Y, Y_M) &= I(Y, Y_M) = \left\langle \log \frac{f(y, y_M)}{f_Y(y) f_{Y_M}(y_M)} \right\rangle = \left\langle \log \frac{f(y|\, y_M)}{f_Y(y)} \right\rangle \\ &= \iint f(y, y_M) \log \frac{f(y|\, y_M)}{f_Y(y)} dy dy_M = \max. \end{aligned} \tag{7.33}$$

If there are reasons for assuming that the joint distribution of the system response and the model response is Gaussian, then

$$I(Y, Y_M) = \frac{1}{2} \ln \left[1 - \rho^2 (Y, Y_M)\right], \tag{7.34}$$

where $\rho(Y, Y_M)$ is the correlation coefficient of Y and Y_M.

The ideas briefly presented in the preceding — including the distance measures between probability distributions and random processes discussed in the two preceding sections — can be useful in the development of a unified approach to stochastic modelling of fatigue. More systematic analysis along this line will be a subject of future research.

39. Applications

Once one develops an appropriate stochastic model which accurately portrays the fatigue crack growth behavior in a structural system under consideration, it can be used as a basis from which reliability estimates can be made. Because of the increasing number of aging engineering structures, one aspect of the reliability problem which has drawn considerable attention is that of fatigue reliability updating through scheduled inspections. We consider the inspection problem in this section to illustrate the use of random fatigue theory in reliability estimation.

To prevent catastrophic failure, fatigue critical structures such as airframes and offshore platforms are usually subjected to scheduled inspection maintenance. For example, a typical fighter aircraft has tens of thousands of fastener holes — each of which serves as a site for crack initiation. The damage tolerance requirement in MIL-A-87221 [211] states that an "airframe shall have adequate residual strength in the presence of flaws for specified periods of usage. These flaws shall be assumed to exist initially in the structure as a result of manufacturing processes." In-service inspections are often essential to meet this requirement.

Determination of optimal inspection and repair strategies requires that the effect of scheduled maintenance on the safety and reliability of the structure be assessed. However, many uncertainties are involved, e.g., initial fatigue quality, in-service load spectrum, material characteristics, crack growth laws, nondestructive evaluation capabilities, etc.

A simple application of the theory of random fatigue to the determination of failure probabilities for structural components under scheduled inspection is presented in the next section. Various uncertainties mentioned previously are taken into account. While the random variable differential equation model for fatigue crack growth is employed in the analysis for simplicity, any of the models discussed in the earlier chapters can be used.

39.1 Reliability Updating through Inspection

Two types of results are considered for an inspection that occurs at time T_i: a) No crack is found during the inspection — implying that the crack size is smaller than the smallest detectable crack size A_{di}, and b) a crack of length A_{mi} is detected during the inspection. These results can be formally represented as

$$A(T_i) < A_{di} \tag{7.35}$$

and

$$A(T_i) = A_{mi}, \tag{7.36}$$

for outcomes (a) and (b), respectively. We note that the minimum detectable crack size is usually random, because a crack longer than A_{di} is undetected with a certain probability, depending on the crack size and inspection method. The distribution of A_{di}, $F_{A_{di}}(a)$, is the distribution of undetected crack size and is identical to the probability of detection curve (POD) for the particular inspection method used. (See Madsen et al. [209].) The measured crack size A_{mi} is generally also random, with distribution function $F_{A_{mi}}(a)$, due to uncertainties in measurement techniques.

If a dominant crack exists in a structure and the failure criterion is based on this crack exceeding a critical crack length, the probability of failure can be written as

$$P_f(t) = P\{A(t) \geq a_c\}, \tag{7.37}$$

where a_c is the critical crack size. Prior to the first inspection, the probability of failure is then

$$P_f(t) = 1 - F_A(a_c;t) = 1 - \int_0^{a_c} f_A(a;t)\, da, \qquad t \leq T_1, \tag{7.38}$$

where

$$f_A(a;t) = \int_0^{\infty} f_A(a;t|x;t_0) f_{A_0}(x)\, dx, \tag{7.39}$$

$f_A(a;t|x;\tau)$ is the probability distribution at time $t > \tau$, conditional on $a(\tau) = x$, and $f_{A_0}(a)$ is the probability distribution of initial crack sizes. Substituting Eq. (7.39) into Eq. (7.38) and integrating, the probability of failure becomes

$$P_f(t) = 1 - \int_0^{\infty} F_A(a_c;t|x;t_0) f_{A_0}(x)\, dx, \tag{7.40}$$

where $F_A(a;t|x;t_0)$ is the cumulative distribution of crack sizes conditional on $A(t_0) = x$.

39.1.1 No Crack Found during First Inspection

If an inspection is conducted at time T_1 and no crack is detected, the updated probability of failure, based on the definition of conditional probability, takes the form,

$$\begin{aligned} P_f^{upd}(t) &= P\{A(t) \geq a_c | A(T_1) < A_{d1}\} \\ &= \frac{P\{A(t) \geq a_c \cap A(T_1) < A_{d1}\}}{P\{A(T_1) < A_{d1}\}}. \end{aligned} \tag{7.41}$$

Equation (7.41) can be written explicitly in terms of the updated distribution of crack sizes as

$$P_f^{upd}(t) = 1 - F_A^{upd}(a_c;t) = 1 - \int_0^{a_c} f_A^{upd}(a;t)\, da, \qquad t > T_1, \tag{7.42}$$

where

$$\begin{aligned} f_A^{upd}(a;t)\, da &= P\{a < A(t) \le a + da | A(T_1) < A_{d_1}\} \\ &= \frac{P\{a < A(t) \le a + da \cap A(T_1) < A_{d1}\}}{P\{A(T_1) < A_{d1}\}}. \end{aligned} \tag{7.43}$$

The numerator in Eq. (7.43) can be expressed as

$$\begin{aligned} &P\{a < A(t) \le a + da \cap A(T_1) < A_{d1}\} \\ &= \int_0^{\infty} P\{a < A(t) \le a + da \cap A(T_1) < A_{d1} \cap A(T_1) = x\}\, dx \\ &= \int_0^{\infty} P\{a < A(t) \le a + da | A(T_1) < A_{d1} \cap A(T_1) = x\} \\ &\qquad \times P\{A(T_1) < A_{d1} | A(T_1) = x\}\ f_A(x;T_1)\, dx, \end{aligned} \tag{7.44}$$

where $f_A(x, T_1)$ is given by Eq. (7.39). We note that

$$\begin{aligned} &P\{a < A(t) \le a + da | A(T_1) < A_{d1} \cap A(T_1) = x\} \\ &\qquad = P\{a < A(t) \le a + da | A(T_1) = x\}. \end{aligned} \tag{7.45}$$

and therefore

$$\begin{aligned} &P\{a < A(t) \le a + da \cap A(T_1) < A_{d1}\} \\ &\qquad = \int_0^{\infty} P\{a < A(t) \le a + da | A(T_1) = x\} \\ &\qquad\qquad \times \{1 - F_{A_{d1}}(x)\}\ f_A(x;T_1)\, dx \\ &\qquad = da \int_0^{\infty} f_A(a;t | x;T_1)\ \{1 - F_{A_{d1}}(x)\} f_A(x;T_1)\, dx. \end{aligned} \tag{7.46}$$

Assuming that A_{d1} is statistically independent of $A(T_1)$, the denominator of Eq. (7.43) can be written as

$$P\{A(T_1) < A_{d1}\} = \int_0^\infty \{1 - F_{A_{d1}}(x)\} f_A(x;T_1)\,dx = k_{d1}. \tag{7.47}$$

Thus, Eq. (7.43) becomes

$$f_A^{upd}(a;t) = \frac{1}{k_{d1}} \int_0^\infty f_A(a;t|x;T_1)\,\{1 - F_{A_{d1}}(x)\} f_A(x;T_1)\,dx, \qquad t > T_1, \tag{7.48}$$

and finally,

$$\begin{aligned} P_f^{upd}(t) &= 1 - \frac{1}{k_{d1}} \int_0^{a_c}\int_0^\infty f_A(a;t|x;T_1)\,\{1 - F_{A_{d1}}(x)\} f_A(x;T_1)\,dx\,da \\ &= 1 - \frac{1}{k_{d1}} \int_0^\infty F_A(a_c;t|x;T_1)\,\{1 - F_{A_{d1}}(x)\} f_A(x;T_1)\,dx, \\ & \qquad t > T_1. \end{aligned} \tag{7.49}$$

39.1.2 Crack Found during First Inspection

Alternatively, when a crack is detected, the updated probability of failure becomes

$$\begin{aligned} P_f^{upd}(t) &= P\{A(t) \geq a_c | A(T_1) = A_{m1}\} \\ &= \frac{P\{A(t) \geq a_c \cap A(T_1) = A_{m1}\}}{P\{A(T_1) = A_{m1}\}}. \end{aligned} \tag{7.50}$$

Following logic similar to that just presented, the updated probability of failure is given by

$$P_f^{upd}(t) = 1 - F_A^{upd}(a_c;t) = 1 - \int_0^{a_c} f_A^{upd}(a;t)\,da, \qquad t > T_1, \tag{7.51}$$

where the updated probability density function of crack sizes is now

$$f_A^{upd}(a;t) = \frac{1}{k_{m1}} \int_0^\infty f_A(a;t|x;T_1) f_{A_{m1}}(x)\, f_A(x;T_1)\,dx, \qquad t > T_1, \tag{7.52}$$

$f_A(x;T_1)$ is given in Eq. (7.39), and

$$k_{m1} = \int_0^{\infty} f_{A_{m1}}(x) f_A(x;T_1)\,dx. \tag{7.53}$$

Substituting Eqs. (7.52) and (7.53) into Eq. (7.51) and integrating, the updated probability distribution becomes

$$P_f^{upd}(t) = 1 - \frac{1}{k_{d1}} \int_0^{\infty} F_A(a_c;t|x;T_1) f_{A_{m1}}(x) f_A(x;T_1)\,dx, \qquad t > T_1. \tag{7.54}$$

39.1.3 No Crack Found during Second Inspection

When no crack is detected during the first inspection, the updated failure probability after the second inspection at a time $T_2 > T_1$ is of interest. If no crack is detected on the second inspection, the updated probability of failure is given by

$$\begin{aligned} P_f^{upd}(t) &= P\{A(t) \geq a_c | A(T_1) < A_{d1} \cap A(T_2) < A_{d2}\} \\ &= \frac{P\{A(t) \geq a_c \cap A(T_1) < A_{d1} \cap A(T_2) < A_{d2}\}}{P\{A(T_1) < A_{d1} \cap A(T_2) < A_{d2}\}}. \end{aligned} \tag{7.55}$$

In terms of the updated distribution of crack sizes, Eq. (7.55) can be written as

$$P_f^{upd}(t) = 1 - F_A^{upd}(a_c;t) = 1 - \int_0^{a_c} f_A^{upd}(a;t)\,da, \qquad t > T_2, \tag{7.56}$$

where we have

$$\begin{aligned} f_A^{upd}(a;t)\,da &= P\{a < A(t) \leq a + da | A(T_1) < A_{d_1} \cap A(T_2) < A_{d2}\} \\ &= \frac{P\{a < A(t) \leq a + da \cap A(T_1) < A_{d1} \cap A(T_2) < A_{d2}\}}{P\{A(T_1) < A_{d1} \cap A(T_2) < A_{d2}\}}, \\ & \qquad t > T_2. \end{aligned} \tag{7.57}$$

The numerator in Eq. (7.43) can be expressed as

$$P\{a < A(t) \le a + da \cap A(T_1) < A_{d1} \cap A(T_2) < A_{d2}\}$$

$$= \int_0^{\infty} P\{a < A(t) \le a + da \cap A(T_1) < A_{d1}$$

$$\cap A(T_2) < A_{d2} \cap A(T_2) = x\}\, dx$$

$$= \int_0^{\infty} P\{a < A(t) \le a + da | A(T_1) < A_{d1} \cap A(T_2) < A_{d2} \cap A(T_2) = x\}$$

$$\times P\{A(T_1) < A_{d1} \cap A(T_2) < A_{d2} \cap A(T_2) = x\}\, dx. \tag{7.58}$$

If we assume that the crack growth process has one step memory, then

$$P\{a < A(t) \le a + da | A(T_1) < A_{d1} \cap A(T_2) < A_{d2} \cap A(T_2) = x\}$$
$$= P\{a < A(t) \le a + da | A(T_2) = x\}, \tag{7.59}$$

since the information at time T_1 will have no bearing on the distribution at time T_2 if we are given the crack size at that time. This assumption is exact for the random variable and the scalar Markov process models, and may be a reasonable approximation for other models. Therefore, Eq. (7.58) becomes

$$P\{a < A(t) \le a + da \cap A(T_1) < A_{d1} \cap A(T_2) < A_{d2}\}$$

$$= k_{d1} \int_0^{\infty} P\{a < A(t) \le a + da | A(T_2) = x\}$$

$$\times P\{A(T_2) < A_{d2} | A(T_1) < A_{d1} \cap A(T_2) = x\} f_A^{upd}(x; T_2)\,\}\, dx$$

$$= da \int_0^{\infty} f_A(a, t | x, T_2)\, \{1 - F_{A_{d2}}(x)\}\, f_A^{upd}(x; T_2)\, dx, \tag{7.60}$$

where $f_A^{upd}(x; T_2)$ is given in Eq. (7.48).

Assuming that A_{d2} is statistically independent of $A(T_2)$, the denominator of Eq. (7.43) can be written as

$$\begin{aligned} &P\{A(T_1) < A_{d1} \cap A(T_2) < A_{d2}\} \\ &\quad = P\{A(T_2) < A_{d2} | A(T_1) < A_{d1}\} P\{A(T_1) < A_{d1}\} \\ &\quad = k_{d1} \int_0^{\infty} \{1 - F_{A_{d2}}(x)\} f_A^{upd}(x;T_2)\, dx = k_{d1} k_{d2}. \end{aligned} \tag{7.61}$$

Thus, Eq. (7.43) becomes

$$f_A^{upd}(a;t) = \frac{1}{k_{d2}} \int_0^{\infty} f_A(a;t| x;T_2) \{1 - F_{A_{d2}}(x)\} f_A^{upd}(x;T_2)\, dx, \qquad t > T_2, \tag{7.62}$$

and finally,

$$\begin{aligned} P_f^{upd}(t) &= 1 - \frac{1}{k_{d2}} \int_0^{a_c} \int_0^{\infty} f_A(a;t| x;T_2) \{1 - F_{A_{d2}}(x)\} f_A^{upd}(x;T_2)\, dx da \\ &= 1 - \frac{1}{k_{d2}} \int_0^{\infty} F_A(a_c;t| x;T_2) \{1 - F_{A_{d2}}(x)\} f_A^{upd}(x;T_2)\, dx, \\ &\qquad t > T_1. \end{aligned} \tag{7.63}$$

39.1.4 Crack Found during Second Inspection

Next, consider the case where a crack is detected during the second inspection. Again, following similar logic as presented earlier, the updated probability of failure can be written,

$$\begin{aligned} P_f^{upd}(t) &= P\{A(t) \ge a_c | A(T_1) \le A_{d1} \cap A(T_2) = A_{m2}\} \\ &= \frac{P\{A(t) \ge a_c \cap A(T_1) \le A_{d1} \cap A(T_2) = A_{m2}\}}{P\{A(T_1) \le A_{d1} \cap A(T_2) = A_{m2}\}}, \end{aligned} \tag{7.64}$$

$$P_f^{upd}(t) = 1 - F_A^{upd}(a_c;t) = 1 - \int_0^{a_c} f_A^{upd}(a;t)\, da, \qquad t \ge T_2, \tag{7.65}$$

where

$$f_A^{upd}(a;t) = \frac{1}{k_{m2}} \int_0^{\infty} f_A(a;t|x;T_2) f_{A_{m2}}(x) f_A^{upd}(x;T_2)\, dx, \qquad t > T_2, \quad (7.66)$$

$f_A^{upd}(x;T_2)$ is given in Eq. (7.48), and

$$k_{m2} = \int_0^{\infty} f_{A_{m2}}(x) f_A^{upd}(x;T_2)\, dx. \quad (7.67)$$

Substituting Eqs. (7.66) and (7.67) into Eq. (7.65) and integrating, the updated probability distribution becomes

$$P_f^{upd}(t) = 1 - \frac{1}{k_{m2}} \int_0^{\infty} F_A(a_c;t|x;T_1) f_{A_{m2}}(x) f_A^{upd}(x;T_2)\, dx, \qquad t > T_2. \quad (7.68)$$

As can be seen, this procedure is recursive and can be repeated for subsequent inspections and combinations of outcomes. If a repair is made, the updated probability of failure is determined based on the initial fatigue quality of the repaired structure. Further details regarding reliability updating through inspection can be found in Madsen [179], Madsen et al. [31], [209], and Yang and Chen [217].

39.2 Numerical Illustration

To illustrate the concepts presented earlier, we consider crack growth damage accumulation in fastener holes of an F–16 lower wing skin. For simplicity, a single crack at a critical location is examined, and the probability of the crack exceeding a critical length is calculated. We note that the entire population of fastener holes should to be examined in a complete analysis.

Consider the random variable crack growth model given by (see Section 30.2)

$$\frac{dA}{dt} = C_0(\gamma) A^{m/2}, \qquad A(t_0) = A_0(\gamma), \quad (7.69)$$

where $C_0(\gamma)$ and $A_0(\gamma)$ are random variables. The probability distribution of the crack size $A(t)$ resulting from this model was derived in Section 30.2 and is repeated here for completeness,

$$F_A(a;t|a_0;t_0) = P\{A(t) \le a\} = \begin{cases} 0, & a \le A_0 \\ \Phi\left(\dfrac{\ln \eta_t - \mu_Z}{\sigma_Z}\right), & a > A_0, \end{cases} \quad (7.70)$$

where $\eta_t = (a_0^{-\beta} - a^{-\beta})/\beta(t - t_0)$, μ_Z and σ_Z are the mean and standard deviation of a normal random variable $Z = \ln C_0$, respectively, and Φ is the standard normal distribution function, and $\beta = (m-2)/2$.

Comparison with experimental data for fatigue cracks in fastener holes [210] indicates that C_0 is lognormally distributed with $\langle \ln C_0 \rangle = -8.91$ and $\sigma_{\ln C_0} = 0.1276$. A probability distribution for $A_0(\gamma)$ that well represents the initial fatigue quality of fastener holes is given by (see Yang and Manning [216])

$$F_{A_0}(a) = \begin{cases} \exp\left[-\left(\dfrac{\ln(a_u/a)}{Q_0\beta_0}\right)^{\alpha_0}\right], & 0 \le a \le a_u \\ 1.0, & a > a_u, \end{cases} \quad (7.71)$$

where $a_u = 0.03$ in. is the upper bound on the initial crack size, and $\alpha_0 = 1.823$ and $Q_0\beta_0 = 1.928$ are shape parameters for the distribution.

Berens and Hovey [203] have employed regression analysis to fit seven different function forms for the probability of detection (POD) curves to the data, which included 22,000 inspections performed on 174 cracks by 107 inspectors. The following model, termed the log-odds–log-scale model, was determined to provide the best fit of the data

$$F_{A_d}(a) = \frac{\exp(\alpha^* + \beta^* \ln a)}{1 + \exp(\alpha^* + \beta^* \ln a)}, \quad 0 \le a \quad (7.72)$$

where α^* and β^* are shape parameters for the distribution. Thus, $F_{A_d}(a)$ is the cumulative distribution of the smallest detectable crack size. Alternatively, it gives the probability of detecting (POD) a crack of length a during an inspection. The parameters in Eq. (7.72) depend upon the inspection interval and the nondestructive evaluation technique. For this example, the parameter values, taken from [217], are $\alpha^* = 13.44$ and $\beta^* = 3.95$. Figure 7.1 depicts the probability of detection curve using these parameters.

One design life for the aircraft is 8,000 flight hours. The effect of inspections for up to two lifetimes, or 16,000 flight hours, will be investigated. The critical crack size a_c is taken to be 0.3 inches. Two inspections will be considered the first at $T_1 = 8{,}000$ flight hours and the second at $T_2 = 12{,}000$ flight hours.

First, we consider the case in which no cracks are found during the first inspection. The distributions of crack size at time $t_0 = 0$ and at time $T_1 = 8{,}000$ flight hours, just before the inspection, are shown in Fig. 7.2. The results in Fig. 7.2 are calculated using Eq. (7.40), where the conditional crack size distribution $F_A(a;t|x;t_0)$ is given by Eq. (7.70). As is seen here, the mean and dispersion of the crack size have increased with time. The updated distribution of crack sizes (i.e., after the first inspection) is calculated via Eq. (7.48) and shown in Fig. 7.3, along with the distribution of crack sizes immediately prior to the inspection, which is given in Fig. 7.2. Finding no cracks during the first inspection has reduced the mean and the dispersion of the crack size, indicating that the probability of the crack exceeding the critical crack size should decrease after the inspection.

If a second inspection is conducted and no crack is found, the updated distribution of crack sizes can be calculated according to Eq. (7.48). Figure 7.4 provides a comparison of the distribution of crack sizes before and after the second inspection, finding no cracks. Again, the mean and dispersion of the crack size is decreased.

Equations (7.40), (7.49), and (7.63) are used to calculate the probability of failure as a function of time, which is shown in Fig. 7.5. Here, we see that the probability that the crack will exceed the critical crack size decreases greatly in the short term and is lower than the probability corresponding to no inspection over the lifetime of the structure.

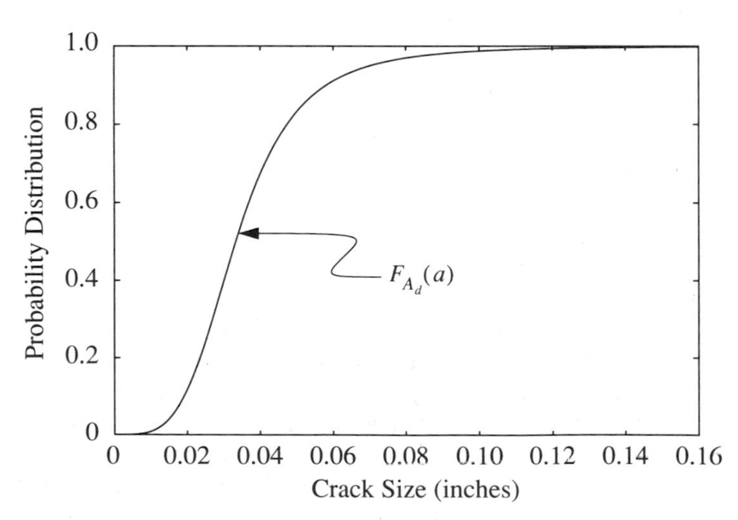

Figure 7.1. Log–odds probability of detection (POD) curve.

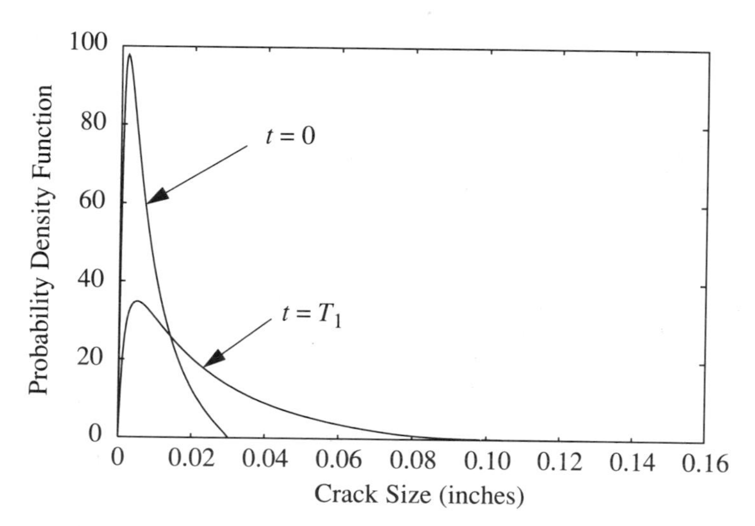

Figure 7.2. Evolution of the crack size probability distribution before the inspection at $T_1 = 8{,}000$ flight hours.

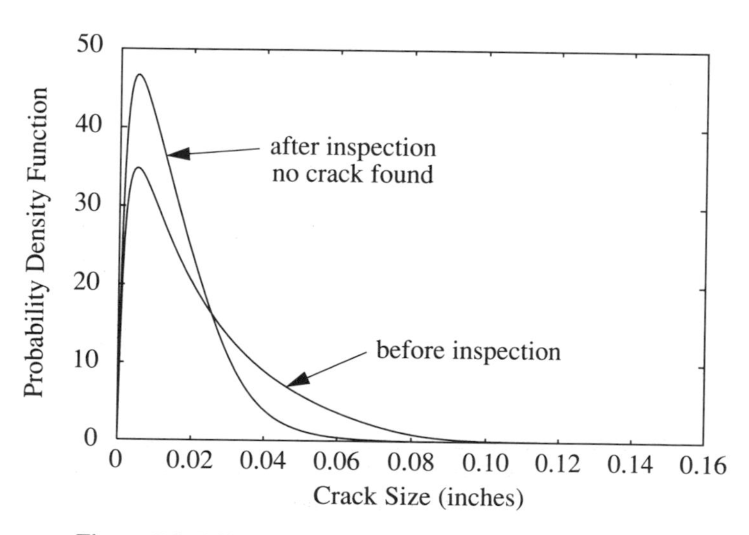

Figure 7.3. Effect on crack size distribution of not finding a crack during the first inspection at $T_1 = 8{,}000$ flight hours.

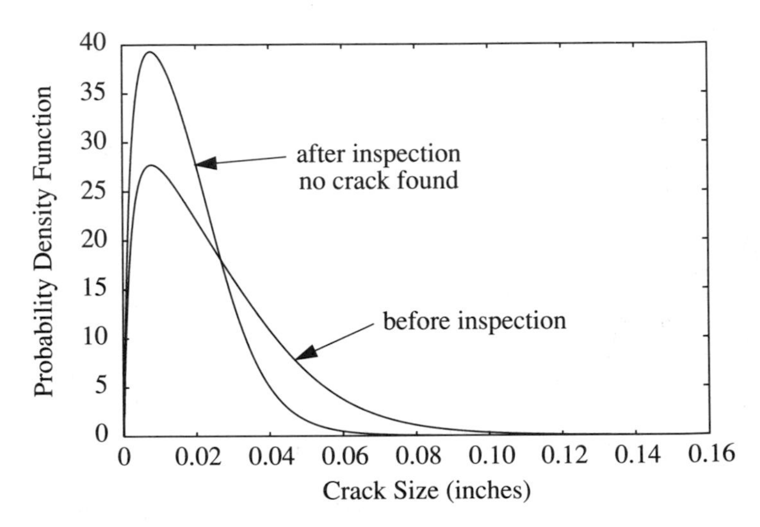

Figure 7.4. Effect on crack size distribution of not finding a crack during the second inspection at $T_2 = 12{,}000$ flight hours; no crack detected during the first inspection.

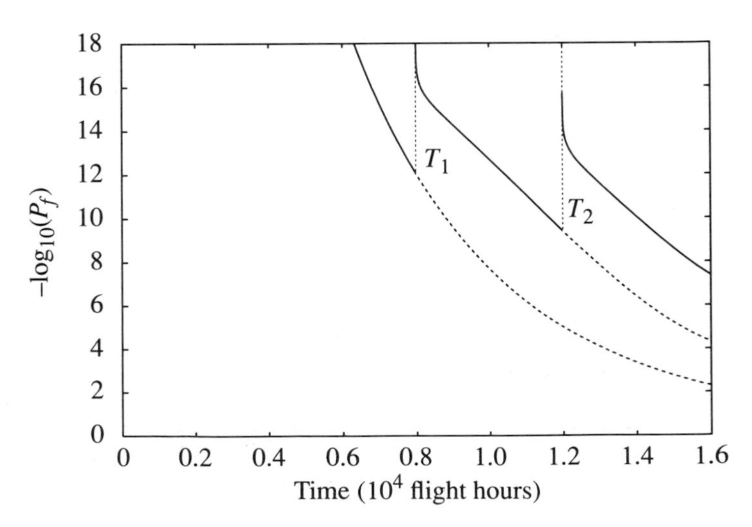

Figure 7.5. Updated probability of failure; no cracks detected during the first and second inspections.

Finally, we consider the case in which no crack is detected during the first inspection, but a crack is detected during the second inspection. The detected crack is assumed to be normally distributed with a mean size of 0.02 in. and a coefficient of variation of 10%, i.e., $F_{A_{m2}}(a)$ in Eq. (7.68) is $N(0.02,0.002)$. The distribution of crack sizes before and after finding the crack on the second inspection, along with the distribution of the measured crack, are shown in Fig. 7.6. Figure 7.7 presents the probability of failure as a function of time for this case. Here, we see that detecting a crack actually decreases the probability of failure for a short time. This is due to the fact that the inspection has confirmed that, even though a crack exists, it has not yet reached the critical crack size. As is seen in Fig. 7.7, the period of time in which the probability of failure decreases is generally short.

Evaluation of the integrals in the preceding equations is possible in only the simplest cases, and for this example, numerical integration was required. As the complexity of the problems and the associated uncertainty increases, direct numerical integration is likely to become impossible. This is especially true when one is looking at extremely small probabilities of failure, which can be inaccurate due to finite precision of the computing hardware and limitations of the computational algorithms. If the uncertainties in the problem and crack growth model are well-represented by random variables, then a computer code such as PROBAN [185] can alleviate some of the numerical difficulties in obtaining the sought failure probabilities. (See also [31] and [209].)

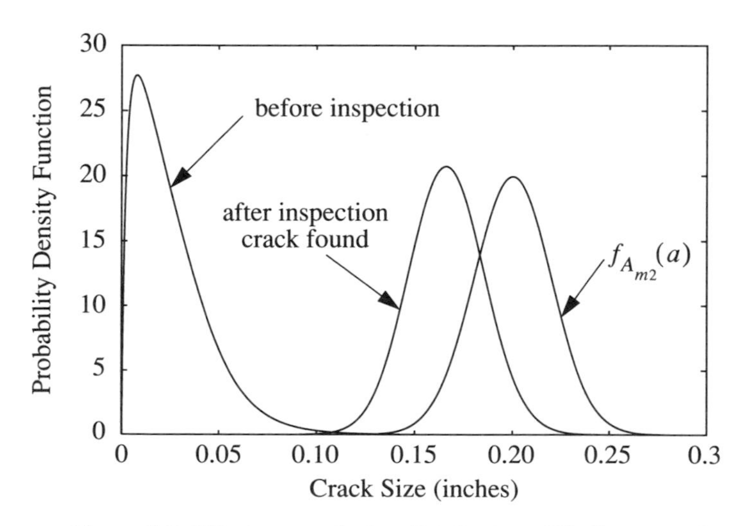

Figure 7.6. Effect on crack size distribution of finding a crack during the second inspection at $T_2 = 12,000$ flight hours; no crack detected during the first inspection.

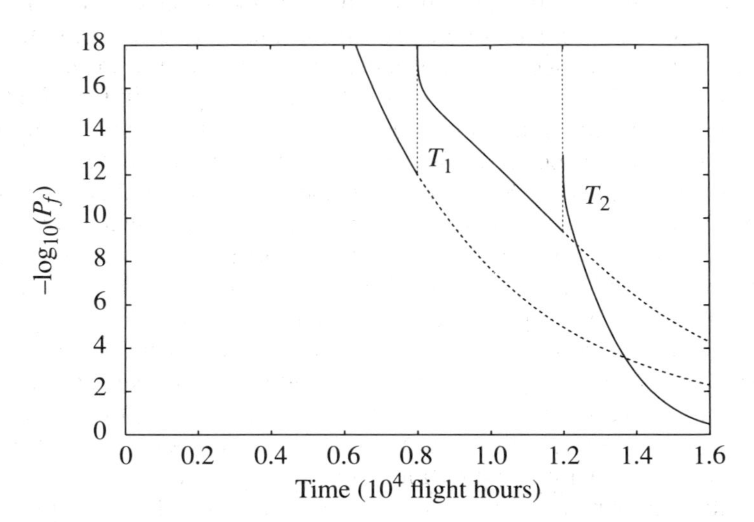

Figure 7.7. Updated probability of failure; no crack detected during first inspection; crack detected during the second inspection.

References

Chapter I

1. ASCE Committee on Fatigue and Fracture Reliability, Series of Articles on Fatigue and Fracture Reliability, *Journal of the Structural Division, ASCE* **108** (ST1), 3-88, (1982).

2. ATLURI, S. N., "Path Independent Integrals in Finite Elasticity and Inelasticity with Body Forces, Inertia and Arbitrary Crack Face Conditions," *Engineering Fracture Mechanics* **16**, 341–364 (1982).

3. BANNANTINE, J. A., J. J. COMER, and J. L. HANDROCK, *Fundamentals of Metal Fatigue Analysis*, Prentice-Hall, Englewood Cliffs, New Jersey, 1990.

4. BEEVERS, C. J. (ed.), *Advances in Crack Length Measurement Techniques*, Chameleon Press, London, 1982.

5. BOLOTIN, V. V., "On the Theory of Corrosion Fatigue," in *Reliability of Mechanics and Structures, Proceedings of the Moskow Energetics Institute* **83**, 5–10 (1986) (in Russian).

6. BOLOTIN, V. V., *Prediction of Service Life for Machines and Structures*, ASME Press, New York, 1989.

7. BRUST, F. W., T. NISHIOKA, S. N. ATLURI, and M. NAKAGAKI, "Further Studies on Elastic-Plastic Stable Fracture Utilizing the T* Integral," *Engineering Fracture Mechanics* **22**, 1079–1103 (1985).

8. CLARKE, G. A., and J. D. LANDES, "Evaluation of J for the Compact Specimen," *Journal of Testing and Evaluation* **7** (5), 264–269 (1979).

9. COLLINS, J. A., *Failure of Materials in Mechanical Design: Analysis, Prediction, Prevention*, John Wiley & Sons, New York, 1981.

10. CORTEN, H. T., and T. J. DOLAN, "Cumulative Fatigue Damage," in *Proceedings of the International Conference on Fatigue of Metals, ASME and IME*, Institution of Mechanical Engineers, London, United Kingdom, pp. 235–246, 1956.

11. CROOKER, T. W., and B. N. LEIS, *Corrosion Fatigue: Mechanics, Metallurgy, Electrochemistry and Engineering*, STP-801, American Society for Testing and Materials, Philadelphia, 1983.

12. DITLEVSEN, O., and R. OLESEN, "Statistical Analysis of the Virkler Data on Fatigue Crack Growth," *Engineering Fracture Mechanics*, **25** (2), 177–195 (1986).

13. DOWLING, N. E., and J. A. BEGLEY, "Fatigue Crack Growth during Gross Plasticity and the J-Integral," in *Mechanics of Crack Growth*, STP-590, American Society for Testing and Materials, Philadelphia, pp. 82–103, 1976.

14. FUCHS, H. O., and R. I. STEPHENS, *Metal Fatigue in Engineering*, John Wiley & Sons, New York, 1980.

15. GHONEM, H., and S. DORE, "Experimental Study of the Constant Probability Crack Growth Curves under Constant Amplitude Loading," *Engineering Fracture Mechanics* **27** (1), 1–25 (1987).

16. GRABACKI, J., "On Continuous Description of the Damage," *Mechanika Teoretyczna i Stosowana* **27** (2), 272–291 (1989).

17. HELLER, R. A. (ed.), *Probabilistic Aspects of Fatigue*, STP-511, American Society for Testing and Materials, Philadelphia, 1972.

18. HENRY, D. L., "Theory of Fatigue Damage Accumulation in Steel," *ASME Transactions* **77**, 913 (1955).

19. HUTCHINSON, J. W., and P. C. PARIS, "Stability Analysis of J Controlled Crack Growth," in *Elastic-Plastic Fracture* (J. D. Landers, J. A. Begley, and G. A. Clarke, eds.) STP-668, American Society for Testing and Materials, Philadelphia, pp. 37–64, 1979.

20. JABLONSKI, D., "An Experimental Study of the Validity of a ΔJ Criterion for Fatigue Crack Growth," in *Nonlinear Fracture Mechanics* (A. Saxena, J. D. Landes, and J. L. Bassani, eds.), American Society for Testing and Materials, Philadelphia, pp. 361–387, 1988.

21. KACHANOV, L. M., "On Creep Rapture Time," *Izviestiia Akademiia Nauk SSSR, Otd. Techn. Nauk* **8**, 26–31 (1958) (in Russian).

22. KACHANOV, L. M., *Introduction to Continuum Damage Mechanics*, Martinus-Nijhoff Publishers, Dordrecht, The Netherlands, 1986.

23. KOCANDA, S., *Fatigue of Metals*, Sijthoff and Noordhoff, Leyden, The Netherlands, 1978.

24. KOZIN, F., and J. L. BOGDANOFF, "Probabilistic Models of Fatigue Crack Growth: Results and Speculations," *Nuclear Engineering and Design* **115**, 143–177 (1989).

25. KRAJCIANOVIĈ, D., "Continuum Damage Mechanics," *Applied Mechanics Review* **37** (1), 1–6 (1984).

26. LECKIE, F. A., "The Constitutive Equations of Continuum Creep Damage Mechanics," *Philosophical Transactions of the Royal Society of London*, Series A, **228**, 27–47 (1978).

27. LEMAITRE, J., "A Continuous Damage Mechanics Model for Ductile Fracture," *Journal of Engineering Materials and Technology* **107**, 83–89 (1985).

28. LEMAITRE, J., and J. L. CHABOCHE, "Aspect Phenomenologique de la Rupture par Endowmagement," *Journal de Mécanique Theorique et Appliqué* **2**, 317–365 (1978).

29. LITTLE, R. E., and J. C. EKVALL, *Statistical Analysis of Fatigue Data*, STP-744, American Society for Testing and Materials, Philadelphia, 1981.

30. MADAYAG, A. F., *Metal Fatigue: Theory and Design*, John Wiley & Sons, New York, 1969.

31. MADSEN, H. O., S. KRENK, and N. C. LIND, *Methods of Structural Safety*, Prentice-Hall, Englewood Cliffs, New Jersey, 1986.

32. MAHMOUD, M. A., and A. HOSSEINI, "Assessment of Stress Intensity Factor and Aspect Ratio Variability of Surface Cracks in Bending Plates," *Engineering Fracture Mechanics* **24** (2), 207–221 (1986).

33. MARCO, S. M., and W. L. STARKEY, "A Concept of Fatigue Damage," *ASME Transactions* **76**, 627 (1954).

34. MILLER, K. J., and M. W. BROWN (eds.), *Multiaxial Fatigue*, STP-853, American Society for Testing and Materials, Philadelphia, 1985.

35. MURAKAMI, S., and N. OHNO, "A Continuum Theory of Creep and Creep Damage," in *Creep in Structures* (A. R. Ponter, and D. R. Hayhurst, eds.), Springer-Verlag, Berlin, pp. 422–444, 1981.

36. PUŜKAR, A., *Microplasticity and Failure of Metallic Materials*, Elsevier, Amsterdam, The Netherlands, 1989.

37. RADON, J. C., S. ARAD, and L. E. CULVER, "Growth of Fatigue Cracks in Metals and Polymers," *Engineering Fracture Mechanics* **6**, 195–208 (1974).

38. RAJU, I. S., and J. C. NEWMAN, "Stress Intensity Factors for a Wide Range of Semi-Elliptical Surface Cracks in Finite-Thickness Plates," *Engineering Fracture Mechanics* **11** (4), 817–829 (1979).

39. RICE, J. R., P. C. PARIS, and J. G. MERKLE, "Some Further Results on J-Integral Analysis and Estimates," in *Fracture Toughness Testing*, STP-536, American Society for Testing and Materials, Philadelphia, pp. 231–245, 1973.

40. ROLFE, S. T., and J. M. BARSOM, *Fracture and Fatigue Control in Structures*, Prentice-Hall, Englewood Cliffs, New Jersey, 1977.

41. SCHIJVE, J., "Observations on the Prediction of Fatigue Crack Growth under Variable Amplitude Loading," in *Fatigue Crack Growth under Spectrum Loading*, STP-595, American Society for Testing and Materials, Philadelphia, pp. 3–23, 1976.

42. SHANG-XIAN, W., "Shape Change of Surface Cracks during Fatigue Growth," *Engineering Fracture Mechanics* **22**, 897–913 (1985).

43. SHANLEY, F. R., "A Theory of Fatigue Based on Unbonding during Reversed Slip," The Rand Corporation, **P-350** (1952).

44. SHIH, C. F., and M. D. GERMAN, "Requirements for a One Parameter Characterization of Crack Tip Fields by the HRR Singularity," *International Journal of Fracture* **17**, 27–43 (1981).

45. SHIH, C. F., and J. W. HUTCHINSON, "Fully Plastic Solutions and Large Scale Yielding Estimates for Plane Stress Crack Problems," *Transactions of ASME, Journal of Engineering Materials and Technology* **98**, 289–295 (1976).

46. SIDOROFF, F., "Description of Anisotropic Damage Application to Elasticity," in *Physical Non-Linearities in Structural Analysis* (J. Hult, and J. Lemaitre, eds.), *Proceedings of the IUTAM Symposium, Senlis, France, 1980*, Springer-Verlag, Berlin, New York, pp. 237-234, 1981.

47. SIH, G. C., "Strain Energy Density Factor Applied to Mixed Mode Crack Problems," *International Journal of Fracture* **10**, 305–321 (1974).

48. SIH, G. C., and P. S. THEORCARIS (eds.), *Mixed Mode Crack Propagation*, Sijthoff, Noordhoff, Alphen ann den Rijn, The Netherlands, 1981.

49. SINCLAIR, G. M., and T. J. DOLAN, "Effect of Stress Amplitude on Statistical Variability in Fatigue Life of 75S-T6 Aluminum Alloy," *Transactions of the American Society of Mechanical Engineers* **75**, 867–872 (1953).

50. SORENSON, A., "A General Theory of Fatigue Damage Accumulation," *Transactions of ASME*, Series D, *Journal of Basic Engineering* **91** (1), 1–14 (1969).

51. STEPHENS, R. I., G. W. MCBURNEY, and L. J. OLIPHANT, "Fatigue Crack Growth with Negative R-Ratio following Tensile Overloads," *International Journal of Fracture Mechanics* **10** (4), 587–592 (1974).

52. TREBULES, V. W., R. ROBERTS, and R. W. HERTZBERG, "Effect of Multiple Overloads on Fatigue Crack Growth in 2024-T3 Aluminum Alloy," *Proceedings of In Flow Growth and Fracture Toughness Testing*, STP-536, American Society for Testing and Materials, Philadelphia, pp. 115–146, 1973.

53. VAKULENKO, A. A., and L. M. KACHANOV, "Continuum Theory of Cracked Media," *Mekhanika Tverdogo Tela* **4**, 159–166 (1971).

54. VIRKLER, D. A., B. M. HILLBERRY, and P. K. GOEL, "The Statistical Nature of Fatigue Crack Propagation," *Journal of Engineering Materials and Technology, ASME* **101**, 148–153 (1979).

Chapter II

55. ANG, A. H-S., and W. H. TANG, *Probability Concepts in Engineering Planning and Design, Volume I – Basic Principles*, John Wiley & Sons, New York, 1975.

56. ANTELMAN, G., and I. R. SAVAGE, "Characteristic Functions of Stochastic Integrals and Reliability Problem," *Naval Research Logistic Quarterly* **12**, 199–122 (1965).

57. BHARUCHA-REID, A. T., *Elements of the Theory of Markov Processes and Their Applications*, McGraw-Hill, New York, 1960.

58. CASTILLO, E., *Extreme Value Theory in Engineering*, Academic Press, New York, 1988.

59. CHHIKARA, R. S., and J. L. FOLKS, "The Inverse Gaussian Distribution as a Life Time Model," *Technometrics* **19**, 461–468 (1977).

60. COX, D. R., and H. D. MILLER, *The Theory of Stochastic Processes*, Chapman and Hall, London, 1977.

61. CRAMER, H., and M. R. LEADBETTER, *Stationary and Related Stochastic Processes*, John Wiley & Sons, New York, 1967.

62. FINE, L. T., *Theories of Probability: An Examination of Foundations*, Academic Press, New York, 1973.

63. GALAMBOS, J., *The Asymptotic Theory of Extreme Order Statistics*, John Wiley & Sons, New York, 1978.

64. GERSCH, W., and R. S-Z. LIU, "Time Series Methods for Synthesis of Random Vibration Systems," *Journal of Applied Mechanics* **43**, 159–164 (1976).

65. GUMBEL, E. J., *Statistics of Extremes*, Columbia University Press, New York, 1958.

66. HARRIS, C. M., and N. SINGPURWALLA, "Life Distributions Derived from Stochastic Hazard Rates," *IEEE Transactions on Reliability* **R-17**, 70–79 (1968).

67. KOZIN, F., "Autoregressive Moving Average Models of Earthquake Records," *Probabilistic Engineering Mechanics* **3** (2), 58–63, (1988).

68. KRENK, S., and J. CLAUSEN, "On the Calibration of ARMA Processes for Simulation" in *Reliability and Optimization of Structural Systems*, (P. Thoft-Christensen, ed.), *Lecture Notes in Engineering*, Springer-Verlag, Berlin, pp. 243–257, 1987.

69. LEADBETTER, M. R., G. LINDGREN, and H. ROOTZEN, *Extremes and Related Properties of Random Sequences and Processes*, Springer-Verlag, New York, 1983.

70. ONICESCU, O., and V. I. ISTRATESCU, "Approximation Theory for Random Functions," *Rendicouti Matematica* **8**, 65–81 (1975).

71. OPPENHEIM, A. V., and R. V. SCHAFER, *Digital Signal Processing*, Prentice-Hall, Englewood Cliffs, New Jersey, 1975.

72. PAKULA, L., "Representation of Stationary Gaussian Process on a Finite Interval," *IEEE Transactions on Information Theory* **IT-20**, 231–232 (1976).

73. SHINOZUKA, M., "Simulation of Multivariate and Multidimensional Random Processes," *Journal of the Acoustical Society of America* **49**, 357–367 (1971).

74. SHINOZUKA, M., "Monte Carlo Solution of Structural Dynamics," *Computers and Structures* **2**, 855–874 (1972).

75. SHUSTER, J. J., "On the Inverse Gaussian Distribution Function," *Journal of American Statistical Association* **63**, 1514–1516 (1968).

76. SOBCZYK, K., "On the Reliability Models for Random Fatigue Damage," in *Proceedings of the 3rd Swedish-Polish Symposium on New Problems in Continuum Mechanics* (O. Brulin, and R. Hsieh, eds.), Waterloo University Press, Waterloo, Ontario, Canada, 1983.

77. SOBCZYK, K., *Stochastic Differential Equations with Application to Physics and Engineering*, Kluwer Academic Publishers, Dordrecht, The Netherlands, 1991.

78. SOONG, T. T., *Probabilistic Modelling and Analysis in Science and Engineering*, John Wiley & Sons, New York, 1981.

79. TWEEDIE, M. C. K., "Statistical Properties of Inverse Gaussian Distribution; Part I," *Annals of Mathematical Statistics* **28**, 362–377, 1957.

80. TWEEDIE, M. C. K., "Statistical properties of inverse Gaussian Distribution; Part II," *Annals of Mathematical Statistics* **28**, 696–705 (1957).

Chapter III

81. ALBELKIS, P. R., and J. M. POTTER (eds.), *Service Fatigue Loads: Monitoring, Simulation, and Analysis*, STP-671, American Society for Testing and Materials, Philadelphia, 1979.

82. BOLOTIN, V. V., *Random Vibrations of Elastic Systems*, Martinus Nijhoff Publishers, Hague, The Netherlands, 1984.

83. CHAKRABARTI, S. K., and R. P. COOLEY, "Statistical Distribution of Periods and Heights of Ocean Waves," *Journal of Geophysical Research* **82**, 1363–1368 (1977).

84. CRANDALL, S. H., and W. D. MARK, *Random Vibration in Mechanical Systems*, Academic Press, New York, 1963.

85. DOWLING, N. E., "Fatigue Failure Predictions for Complicated Stress-Strain Histories," *Journal of Materials* **7** (1), 71–87 (1972).

86. DOWNING, S. D., and D. F. SOCIE, "Simplified Rainflow Counting Algorithms," *International Journal of Fatigue* **4** (1), 31–40 (1982).

87. KAREEM, A., "Wind Effects on Structures: A Probabilistic Viewpoint," *Probabilistic Engineering Mechanics* **2** (4), 166–100 (1987).

88. KRENK, S., *A Double Envelope for Stochastic Processes*, Report No. 134, The Danish Center for Applied Mathematics and Mechanics, Lyngby, Denmark, 1978.

89. KRENK, S., H. O. MADSEN, and P. H. MADSEN, "Stationary and Transient Response Envelopes," *Journal of Engineering Mechanics, ASCE* **109**, 263–278 (1983).

90. KRENK, S., and H. GLUVER, *A Markov Matrix for Fatigue Load Simulation and Rain-Flow Range Evaluation*, Report No. 388, The Danish Center for Applied Mathematics and Mechanics, Lyngby, Denmark, 1989.

91. LINDGREN, G., "Wave-Length and Amplitude in Gaussian Noise," *Advances in Applied Probability* **4**, 81–108 (1972).

92. LINDGREN, G., and I. RYCHLIK, "Wave Characteristic Distributions for Gaussian Waves — Wave-Length, Amplitude and Steepness," *Ocean Engineering* **9** (5), 411–432 (1982).

93. LINDGREN, G., and I. RYCHLIK, "Rain Flow Cycle Distributions for Fatigue Life Prediction Under Gaussian Load Processes," *Fatigue of Engineering Materials and Structures* **10** (3), 251–260 (1987).

94. LONGUET-HIGGINS, M. S., "On the Joint Distribution of Wave Periods and Amplitudes in Random Wave Field," *Proceedings of the Royal Society of London* **A389**, 241–258 (1983).

95. LUTES, L. D., M. CORAZAO, J. S. HU, and J. ZIMMERMAN, "Stochastic Fatigue Damage Accumulation," *Journal of Structural Engineering, ASCE* **110** (11), 2585–2601 (1984).

96. NAESS, A., "Prediction of Extremes of Morison-Type Loading; An Example of a General Method," *Ocean Engineering* **10**, 313–324 (1983).

97. OCHI, M., "Non-Gaussian Random Processes in Ocean Engineering," *Probabilistic Engineering Mechanics* **1** (1), 28–39 (1986).

98. ORTIZ, K., "On the Stochastic Modelling of Fatigue Crack Growth," Ph.D. Thesis, Stanford University, Stanford, California, 1985.

99. POTTER, J. M., and R. T. WATANABE (eds.), *Development of Fatigue Loading Spectra*, STP-1006, American Society for Testing and Materials, Philadelphia, 1989.

100. RICE, S. O., "Mathematical Analysis of Random Noise," *Bell Systems Technical Journal* **23**, 282–332 (1944); **24**, 46–156 (1945); also in *Selected Papers on Noise and Stochastic Processes* (N. Wax, ed.), Dover, New York, 1954.

101. RICE, J. R., and F. P. BEER, "On the Distribution of Rises and Falls in a Continuous Random Process," *Journal of Basic Engineering* **87**, 398–404 (1965).

102. RYCHLIK, I., "A New Definition of the Rain-Flow Cycle Counting Method," *International Journal of Fatigue* **9**, 119–121 (1987).

103. RYCHLIK, I., "Rain-Flow Cycle Distribution for Ergodic Load Processes," *SIAM Journal of Applied Mathematics* **48**, 662–679 (1988).

104. RYCHLIK, I., "Simple Approximations of the Rain-Flow Cycle Distribution for Discretized Random Loads," *Probabilistic Engineering Mechanics* **4**, 40–48 (1989).

105. SOBCZYK, K., and D. B. MACVEAN, "Non-Stationary Random Vibrations of Systems Travelling with Variable Velocity," in *Stochastic Problems in Dynamics* (B. D. Clarkson, ed.), Pitman, London, 1977.

106. SIMIU, E., and R. H. SCANLAN, *Wind Effects on Structures: An Introduction to Wind Engineering*, John Wiley & Sons, New York, 1966.

107. TUNG, C. C., "Peak Distribution of Random Wave-Current Forces," *Journal of the Engineering Mechanics Division, ASCE* **100** (EM5), 875–889 (1974).

108. WINTERSTEIN, S. R., *Diffusion Models and the Energy Fluctuation Scale: A Unified Approach to Extremes and Fatigue*, Technical Report No. 64, Department of Civil Engineering, Stanford University, Stanford, California, 1984.

109. WIRSCHING, P. H., and A. M. SHEHATA, "Fatigue Under Wide Band Random Stresses Using the Rain-Flow Method," *Journal of Engineering Materials and Technology, ASME*, **99** 205–211 (1977).

110. YANG, J. N., "Nonstationary Envelope Process and First Excursion Probability," *Journal of Structural Mechanics* **1**, 231–248 (1972).

Chapter IV

111. BOGDANOFF, J. L., "A New Cumulative Damage Model, Part I," *Journal of Applied Mechanics* **45**, 245–250 (1978).

112. BOGDANOFF, J. L., and W. KRIEGER, "A New Cumulative Damage Model, Part II," *Journal of Applied Mechanics* **45**, 251–257 (1978).

113. BOGDANOFF, J. L., "A New Cumulative Damage Model, Part III," *Journal of Applied Mechanics* **45**, 733–739 (1978).

114. BOGDANOFF, J. L., and F. KOZIN, "A New Cumulative Damage Model, Part IV," *Journal of Applied Mechanics* **47**, 40–44 (1980).

115. BOGDANOFF, J. L., and F. KOZIN, *Probabilistic Models of Cumulative Damage*, John Wiley & Sons, New York, 1985.

116. COX, B. N., and W. L. MORRIS, "A Probabilistic Model of Short Fatigue Crack Growth," *Fatigue and Fracture of Engineering Materials and Structures* **10** (5), 419–428 (1987).

117. COX, B. N., and W. L. MORRIS, "Model-Based Statistical Analysis of Short Fatigue Crack Growth in Ti6A1-25n-4Zr-6Mo," *Fatigue and Fracture of Engineering Materials and Structures* **10** (6), 429–446 (1987).

118. FELLER, W., *Introduction to Probability Theory and Its Applications*, John Wiley & Sons, New York, 1968.

119. GHONEM, H., and J. W. PROVAN, "The Micromechanics Theory of Fatigue Crack Initiation and Propagation," *Engineering Fracture Mechanics* **13**, 963–77 (1980).

120. JAMES, M. R., and W. L. MORRIS, "Effect of Fracture Surface Roughness on Growth of Short Fatigue Cracks," *Metallurgical Transactions* **A14**, 153–155 (1983).

121. KEIDING, N., "Estimation in Birth Processes," *Biometrika* **61**, 71–80 (1974).

122. MISRA, P. N., and H. W. SORENSON, "Parameter Estimation in Poisson Processes," *IEEE Transactions on Information Theory* **21**, 87–90 (1975).

123. MORGESEN, H., *The Nature of Physical Reality*, McGraw-Hill, New York, 1950.

124. NEMEC, J., and J. SEDLACEK, *Statistical Problems of Safety of Structures*, Academia, Prague, Czechoslovakia, 1982 (in Czech).

125. OH, K. P., "A Diffusion Model for Fatigue Crack Growth," *Proceedings of Royal Society of London* **A367**, 47–58 (1979).

126. OH, K. P., "A Weakest Link Model for the Prediction of Fatigue Crack Growth Rate," *Journal of Engineering Materials and Technology* **100**, 170–174 (1978).

127. SOBCZYK, K., "On the Markovian Models for Fatigue Accumulation," *Journal de Mécanique Theorique et Appliqué*, (Numor Special), 147–160 (1982).

128. SOBCZYK, K., "Stochastic Models for Fatigue Damage of Materials," *Advances of Applied Probability* **19**, 652–673 (1987).

Chapter V

129. ABRAMOWITZ, M., and I. A. STEGUN (eds.), *Handbook of Mathematical Functions*, Dover, New York, 1970.

130. ARONE, R., "Fatigue Crack Growth under Random Overloads Superimposed on Constant-Amplitude Cyclic Loading," *Engineering Fracture Mechanics* **24** (2), 223–232 (1986).

131. BELL, P. D., and A. WOLFMAN, "Mathematical Modelling of Crack Growth Interaction Effects," in *Fatigue Crack Growth under Spectrum Loads*, STP-595, American Society for Testing and Materials, Philadelphia, pp. 157–171, 1976.

132. BROEK, D., and S. H. SMITH, "The Prediction of Fatigue Crack Growth under Flight-by-Flight Loading," *Engineering Fracture Mechanics* **11**, 123–141 (1979).

133. DITLEVSEN, O., and K. SOBCZYK, "Random Fatigue Crack Growth with Retardation," *Engineering Fracture Mechanics* **24** (6), 861–878 (1986).

134. ITAGAKI, H., and M. SHINOZUKA, "Application of the Monte Carlo Technique to Fatigue-Failure Analysis under Random Loading," in *Probabilistic Aspects of Fatigue* (R. A. Heller, ed.), STP-511, American Society for Testing and Materials, Philadelphia, pp. 168-184, 1972.

135. JONES, R. E., "Fatigue Crack Growth Retardation after Single-Cycle Peak Overload in Ti 611-4V Titanium Alloy," *Engineering Fracture Mechanics* **5**, 585–604 (1973).

136. KOCANDA, S., and J. SZALA, *Foundations of Fatigue Calculations*, Polish Scientific Publications, Warsaw, Poland, 1985 (in Polish).

137. KOCANDA, S., and J. SADOWSKI, "Correlational Study of Fatigue Crack Growth Rate in Steel Elements of Improved Strength," *Archiwum Budowy Maszyn* **24** (1977) (in Polish).

138. KOGAJEV, V. H., and S. H. LIEBIEDINSKIJ, "Probabilistic Model for Fatigue Crack Growth," *Mashinoviedienije* **4**, 78–83 (1983) (in Russian).

139. KOTZ, S., "Multivariate Distribution at a Cross Road," in *Statistical Distributions in Scientific Work, Vol. 1* (C. Taillie, G. P. Patil, and B. A. Baldessari, eds.), Reidel Publishing Company, Dordrecht, The Netherlands, pp. 245–270, 1975.

140. KOZIN, F., and J. L. BOGDANOFF, "On Probabilistic Modelling of Fatigue Crack Growth," in *Structural Safety and Reliability, Vol. 2, Proceedings of ICOSSAR '85, Kobe, Japan* (I. Konishi, A. H-S. Ang, and M. Shinozuka, eds.), Elsevier, Amsterdam, The Netherlands, pp. 331–340, 1985.

141. LIU, P. L., and A. DER KIUREGHIAN, "Multivariate Distribution Models with Prescribed Marginals and Covariances," *Probabilistic Engineering Mechanics* **1** (2), 105–112 (1986).

142. MORGENSTERN, D., "Einfache Beispiele Zweidimensionaler Verteilungen," *Mitteilungsblatt fur Mathematische Statistik* **8**, 234–235 (1956).

143. PORTER, T. R., "Method of Analysis and Prediction for Variable Amplitude Fatigue Crack Growth," *Engineering Fracture Mechanics* 4, 717–736 (1972).

144. ROBIN, C., M. LOUAH, and G. PLUVINAGE, "Influence of an Overload on the Fatigue Crack Growth in Steels," *Fatigue Engineering Materials and Structures* **6**, 1–13 (1983).

145. SCHÜTZ, W., "The Prediction of Fatigue Life in the Crack Initiation and Propagation Stages — a State of the Art Survey," *Engineering Fracture Mechanics* **11**, 405–421 (1979).

146. SCHIJVE, J., "Four Lectures on Fatigue Crack Growth," *Engineering Fracture Mechanics* **11**, 168–221 (1979).

147. SNYDER, D. L., *Random Point Processes,* John Wiley & Sons, New York, 1975.

148. SOBCZYK, K., "Probabilistic Modelling of Fatigue Crack Growth under Variable-Amplitude Loading," in *Random Vibration — Status and Recent Developments* (I. Elishakoff, and R. H. Lyon, eds.), Elsevier, Amsterdam, The Netherlands, pp. 451–458, 1987.

149. SOBCZYK, K., and J. TREBICKI, "Modelling of Random Fatigue by Cumulative Jump Processes," *Engineering Fracture Mechanics* **34** (2), 477–493 (1989).

150. SOBCZYK, K., and J. TREBICKI, "Cumulative Jump-Correlated Model for Random Fatigue," *Engineering Fracture Mechanics*, in press.

151. SOBCZYK, K., and J. TREBICKI, "Maximum Entropy Principle in Stochastic Dynamics," *Probabilistic Engineering Mechanics* **5** (3) (1990).

152. SPENCER, B. F., JR., and L. A. BERGMAN, "On the Estimation of Failure Probability having Prescribed Statistical Moments of First Passage Time," *Probabilistic Engineering Mechanics* **1** (3) (1986).

153. STOUFFER, D. C., and J. F. WILLIAMS, "A Method for Fatigue Crack Growth with a Variable Stress Intensity Factor," *Engineering Fracture Mechanics* **11**, 525–536 (1979).

154. WEI, R. P., "Fracture Mechanics Approach to Fatigue Analysis and Design," *Journal Engineering Materials and Technology* **100**, 113–120 (1978).

155. WEI, R. P., and T. T. SHIH, "Delay in Fatigue Crack Growth," *International Journal Fracture* **10**, 77–85 (1974).

156. WINTERSTEIN, S. R., and P. S. VEERS, "Diffusion Models of Crack Growth due to Random Overloads," in *Probabilistic Methods in Civil Engineering* (P. D. Spanos, ed.), American Society of Civil Engineers, New York, pp. 508–511, 1988.

Chapter VI

157. AKITA, S., and M. ICHIKAWA, "A New Model for Scatter of Fatigue Crack Growth Rate," *International Journal Fracture* **32**, R3–R6 (1986).

158. ASTM, "Appendix I, E647-86a, Recommended Data Reduction Techniques," in *Annual Book of ASTM Standards. Vol. 3.01: Metals Test Methods and Analytical Procedures*, American Society for Testing and Materials, Philadelphia, pp. 919–920, 1987.

159. BARSOM, J. M., "Fatigue-Crack Growth under Variable-Amplitude Loading in ASTM A-514-B Steel," in *Progress in Flaw Growth and Fracture Testing*, STP-536, American Society for Testing and Materials, Philadelphia, pp. 147–167, 1973.

160. BOLOTIN, V. V., "On Safe Crack Size under Random Loading," *Izvestiia Akademiia Nauk SSSR, Mekhanika Tverdogo Tela* **1**, (1980). (in Russian).

161. BOLOTIN, V. V., "Lifetime Distribution under Random Loading," *Zhurnal Priklandnoi Mekhaniki, Tekhnicheskoi Fiziki* **5**, (1980) (in Russian).

162. DITLEVSEN, O., "Random Fatigue Crack Growth — A First Passage Problem," *Engineering Fracture Mechanics* **23** (2), 467–477 (1986).

163. DITLEVSEN, O., and R. OLESEN, "Statistical Analysis of the Virkler Data on Fatigue Crack Growth," *Engineering Fracture Mechanics* **25** (2), 177–195 (1986).

164. DOLINSKI, K., "Stochastic Loading and Material Inhomogeneity in Fatigue Crack Propagation," *Engineering Fracture Mechanics* **25**, 809–818 (1986).

165. DOLINSKI, K., "Stochastic Modelling and Statistical Verification of Crack Growth under Constant Amplitude Loading," (manuscript) (1991).

166. DOVER, W. D., and R. D. HIBBERT, "The Influence of Mean Stress and Amplitude Distribution on Random Load Fatigue Crack Growth," *Engineering Fracture Mechanics* **9**, 251–263 (1977).

167. ENNEKING,T. J., "On the Stochastic Fatigue Crack Growth Problem," Ph.D. Dissertation, Department of Civil Engineering, University of Notre Dame, Notre Dame, Indiana, 1991.

168. ENNEKING, T. J., B. F. SPENCER, JR., and I. P. E. KINNMARK, "Stationary Two-State Variable Problems in Stochastic Mechanics," *Journal of Engineering Mechanics, ASCE* **116** (2), 334–358 (1990).

169. FERGUSON, R. I., "River Loads Underestimated by Rating Curves," *Water Resources Research* **22** (1), 74–76 (1986).

170. FICHERA, G., "On a Unified Theory of Boundary Value Problems for Elliptic-Parabolic Equations of Second Order," in *Boundary Problems in Differential Equations* (R. E. Langer, ed.), University of Wisconsin Press, Madison, Wisconsin, pp. 97–102, 1960.

171. GUERS, F., and R. RACKWITZ, "Time-Variant Reliability of Structural Systems subject to Fatigue," in *Proceedings of ICASP-5, Vol. 1, Vancouver, Canada,* pp. 497–505, 1987.

172. IMSL, *The International Math and Statistics Subroutine Library,* IMSL, Inc., Houston, Texas, 1984.

173. JOHNSON, W. S., "Multi-Parameter Yield Zone Model for Predicting Spectrum Crack Growth," in *Methods and Models for Predicting Fatigue crack Growth under Random Loading* (J. B. Chang, and C. M. Hudson, eds.), STP-748, American Society for Testing and Materials, Philadelphia, pp. 85–102, 1981.

174. KOZIN, F., and J. L. BODGANOFF, "A Critical Analysis of some Probabilistic Models of Fatigue Crack Growth," *Engineering Fracture Mechanics* **14** (1), 59–89 (1981).

175. KUNG, C. J., and K. ORTIZ, "Objective Comparison of Fatigue Crack Growth Laws," *in Structural Safety and Reliability, Vol. 2* (A. H-S. Ang, M. Shinozuka, and G. I. Schuëller, eds.) American Society of Civil Engineers, New York, pp. 1627–1630, 1990.

176. LIN, Y. K., and J. N. YANG, "On Statistical Moments of Fatigue Crack Propagation," *Engineering Fracture Mechanics* **18** (2), 243–256 (1983).

177. LIN, Y. K., and J. N. YANG, "A Stochastic Theory of Fatigue Crack Propagation," *AIAA Journal* **23** (1), 117–124 (1985).

178. MADSEN, H. O., "Deterministic and Probabilistic Models for Damage Accumulation due to Time Varying Loading," *DIALOG 5-82*, Danish Engineering Academy, Lyngby, Denmark, 1983.

179. MADSEN, H. O., "Random Fatigue Crack Growth and Inspection," in *Structural Safety and Reliability, Vol. 1, Proceedings of ICOSSAR '85, Kobe, Japan* (I. Konishi, A. H-S. Ang, and M. Shinozuka, eds.), Elsevier, Amsterdam, The Netherlands, pp. 475–484 (1985).

180. MEHR, C. B., and J. A. MCFADDEN, "Certain Properties of Gaussian Processes and Their First-Passage Times," *Journal of Royal Statistical Society*, Series B, **27**, 505–522 (1965).

181. NEWBY, M., "Likelihood Methods and the Analysis of Fatigue Crack Growth," *Engineering Fracture Mechanics* **37** (3), 701–705 (1990).

182. ORTIZ, K., and A. S. KIREMIDJIAN, "Time Series Analysis of Fatigue Crack Growth Data," *Engineering Fracture Mechanics* **24** (5), 657–676 (1986).

183. ORTIZ, K., and A. S. KIREMIDJIAN, "Stochastic Modeling of Fatigue Crack Growth," *Engineering Fracture Mechanics* **29** (3), 657–676 (1988).

184. OSTERGAARD, D. F., and B. M. HILLBERRY, "Characterization of Variability in Fatigue Crack Propagation Data," in *Probabilistic Methods for Design and Maintenance of Structures,* STP-798, American Society for Testing and Materials, Philadelphia, pp. 97-115, 1983.

185. PROBAN-2, A.S. Veritas Research, Det norske Veritas, Oslo, Norway, 1989.

186. PROVAN, J. (ed.) *Probabilistic Fracture Mechanics and Reliability,* Martinus Nijhoff Publishers, Dordrecht, The Netherlands, 1987.

187. SOBCZYK, K., "Stochastic Modeling of Fatigue Crack Growth," in *Proceedings of the IUTAM Symposium on 'Probabilistic Methods in Mechanics of Solids and Structures,' Stockholm, Sweden*, Springer-Verlag, Berlin, pp. 111–119, 1984.

188. SOBCZYK, K., "Modelling of Random Fatigue Crack Growth," *Engineering Fracture Mechanics* **24** (4), 609–623 (1986).

189. SOLOMOS, G. P., "First-Passage Solutions in Fatigue Crack Propagation," *Probabilistic Engineering Mechanics* **4** (1), 32–39 (1989).

190. SPENCER, B. F., JR., and J. TANG, "A Markov Process Model for Fatigue Crack Growth," *Journal of Engineering Mechanics, ASCE* **114** (12), 2134–2157 (1988).

191. SPENCER, B. F., JR., J. TANG, and M. E. ARTLEY, "A Stochastic Approach to Modeling Fatigue Crack Growth," *Journal of the AIAA* **27** (11), 1628–1635 (1989).

192. SWANSON, S. R., "Random Load Fatigue; a State of the Art Survey," *Materials Research Standards MTRSA* **8** (4), 10–44 (1968).

193. TANAKA, H., and A. TSURUI, "Random Propagation of a Semi-Elliptical Surface Crack as a Bivariate Stochastic Process," *Engineering Fracture Mechanics* **33** (5), 787–800 (1989).

194. TANG, J., and B. F. SPENCER, JR., "Reliability Solution for the Stochastic Fatigue Crack Growth Problem," *Engineering Fracture Mechanics* **34** (2), 419–433 (1989).

195. TSURUI, A., and H. ISHIKAWA, "Application of Fokker–Planck Equation to a Stochastic Fatigue Crack Growth Model," *Structural Safety* **4**, 15–29 (1986).

196. TSURUI, A., J. NIENSTEDT, G.I. SCHUËLLER, and H. TANAKA, "Time-Variant Structural Reliability using Diffusive Crack Growth Models," *Engineering Fracture Mechanics* **34** (1), 153–167 (1989).

197. VEERS, P. J., "Fatigue Crack Growth due to Random Loading," Ph.D. Dissertation, Department of Mechanical Engineering, Stanford University, Stanford, California, 1987.

198. VEERS, P. J., S. R. WINTERSTEIN, D. V. NELSON, and C. A. CORNELL, "Variable Amplitude Load Models for Fatigue Damage and Crack Growth," in *Development of Fatigue Loading Spectra,* STP-1006, American Society for Testing and Materials, Philadelphia, pp. 172–197, 1989.

199. WINTERSTEIN, S. R., and P. S. VEERS, "Diffusion Models of Fatigue Crack Growth with Sequence Effects due to Stationary Random Loads," *Structural Safety and Reliability, Vol. 2* (A. H-S. Ang, M. Shinozuka, and G. I. Schuëller, eds.), American Society of Civil Engineers, New York, pp. 1523–1530, 1990.

200. YANG, J. N., G. C. SALIVAR, and C. G. ANNIS, "Statistical Modeling of Fatigue-Crack Growth in a Nickel-Based Superalloy," *Engineering Fracture Mechanics* **18** (2), 257–270 (1983).

201. YANG, J. N., and R. C. DONATH, "Statistical Crack Propagation in Fastener Holes under Spectrum Loading," *Journal of Aircraft* **20** (12), 1028–1032 (1983).

202. YANG, J. N., W. H. HSI, and S. D. MANNING, "Stochastic Crack Propagation with Application to Durability and Damage Tolerance Analyses," in *Probabilistic Fracture Mechanics and Reliability* (J. Provan, ed.) Martinus Nijhoff Publishers, The Netherlands, 1987.

Chapter VII

203. BERENS, A. P., and P. W. HOVEY, "Evaluation of NDE Reliability Characterization," Report No. AFWAL-TR-81-4160, Volume I, University of Dayton Research Institute, Dayton, Ohio, 1981.

204. CSISZAR, I., "Information-Type Distance Measure and Indirect Observations," *Studia Scientiarum Mathematicarum Hungarica* **2**, 299–318 (1967).

205. GOODWIN G. C., and R. I. PAYNE, *Dynamic System Identification; Experiment Design and Data Analysis*, Academic Press, New York, 1977.

206. KARAKOS, D., and P. PAPONTONI-KAZAKOS, "Spectral Distance Measures between Gaussian Processes," *IEEE Transactions on Automatic Control* **AC-25**, 950–959 (1980).

207. KERRIDGE, D. F., "Inaccuracy and Inference," *Journal of the Royal Statistical Society*, Series B, **23**, 184–195 (1961).

208. KULLBACK, S., *Information Theory and Statistics*, John Wiley & Sons, New York, 1959.

209. MADSEN, H. O., R. SKJONG, A. G. TALLIN, and F. KIRKEMO, "Probabilistic Fatigue Crack Growth Analysis of Offshore Structures with Reliability Updating through Inspection," in *Proceedings of the Marine Structural Reliability Symposium, Arlington, Virginia*, pp. 45-55, 1987.

210. MANNING, S. D., and J. N. YANG, "USAF Durability Design Handbook: Guidelines for the Analysis and Design of Durable Aircraft Structures," Technical Report AFFDL-TR-84-3027, Wright–Patterson Air Force Base, Ohio, February, 1984.

211. MIL-A-87221, "General Specifications for Aircraft Structures," U. S. Air Force Aeronautical Systems Division, Wright-Patterson Air Force Base, Ohio, 1985.

212. POPLAVSKY, R. P., "Thermodynamical Models of Information Processes," *Uspiekhi Fiziceckikh Nauk* **115** (3) (1975) (in Russian).

213. SOBCZYK, K., "Theoretic Information Approach to Identification and Signal Processing," *Proceedings of IFIP Conference on Reliability and Optimization of Structural Systems* (ed. Thoft-Christensen, ed.), *Lecture Notes in Engineering* **33**, Springer-Verlag, Berlin, 1987.

214. SUGIMOTO, S., and T. WADA, "Spectral Expressions of Information Measures of Gaussian Time Series and Their Relation to AIC and CAT," *IEEE Transactions on Information Theory* **34** (4) (1988).

215. WHITTLE, P., "The Analysis of Multiple Stationary Time Series," *Journal Royal Statistical Society*, Series B, **15**, 125–139 (1953).

216. YANG, J. N., and S. D. MANNING, "Distribution of Equivalent Initial Flaw Size," *Proceedings of the Reliability and Maintainability Conference*, 112–120 (1980).

217. YANG, J. N., and S. CHEN, "Fatigue Reliability of Structural Components under Scheduled Inspection and Repair Maintenance," *Probabilistic Methods in Mechanics of Solids and Structures, Proceedings of the IUTAM Symposium, Stockholm, 1984,* Springer-Verlag, Berlin, pp. 559–568, 1985.

Author Index

D

E

F

G

H

I

J

K

L

M

N

O

P

R

S

T

V

W

Y

Z

Subject Index